THE SYMBIOTIC UNIVERSE

Also by George Greenstein

FROZEN STAR (1984)

THE SYMBIOTIC UNIVERSE

LIFE AND MIND IN THE COSMOS

George Greenstein

Illustrations by Dolores R. Santoliquido

WILLIAM MORROW AND COMPANY, INC.
NEW YORK

Library of Congress Cataloging-in-Publication Data

Greenstein, George, 1940–
    The symbiotic universe: life and mind in the cosmos/
George Greenstein.
    p.   cm.
    ISBN 0-688-07604-1
    1. Cosmology. 2. Life—Origin. I. Title.
QB981.G746 1988
523.1—dc19   87-17391
                              CIP

Printed in the United States of America

First Edition

1 2 3 4 5 6 7 8 9 10

BOOK DESIGN BY PANDORA SPELIOS/MARIA EPES

To F.C.

ACKNOWLEDGMENTS

"It is remarkable how much of the effort is conducted in conversations," I wrote in my first book, describing the struggle to understand the nature of pulsars. Little did I know the same would apply to me. Never could I have come to what little comprehension I possess of the strange subject of the present book without the generous assistance of others. Alan Babb, Francis Couvares, Stephen George, Alan Guth, Kannan Jagannathan, Lisa Raskin, Ronald Tiersky, Paul Vitz, and Arthur Zajonc—some gave of their specialized knowledge, others of much-valued advice and counsel, yet others helped me to hammer out the implications of my ideas. Terry Allen and Theron Raines gave invaluable editorial assistance. Nothing could have been done without them; to each my deepest thanks.

Contents

PART TWO: Mind

CONTENTS

Prologue

The Second Sun

The Second Sun came in out of the constellation of Lyra at twenty thousand miles per hour. It was surrounded by an immense number of comets riding out ahead at vast distances, far more numerous than those surrounding our own sun. They scattered across our skies like luminescent feathers in those years, their plumes streaming outward. Each night several could be seen. As for the star, it was the brightest in the sky and shone a brilliant red. There was something almost iridescent about its gleam; people named it Lucifer.

Through telescopes Lucifer was resolved into a disk, its face disfigured by immense dark patches, similar to sunspots but far larger. "Starspots," astronomers called them. The star had prominences too, vast flamelike structures arching high above its surface—again like the Sun's, but again vaster and more numerous.

Lucifer grew brighter as the millennia passed. It came to

dominate the sky. Popular songs were written about it, bad novels and sentimental poems. Lovers plighted their troth on hilltops by its great light. During the final century it was so brilliant you could see the thing in broad daylight. That was in the wintertime. Summers it was up at night, and drowned out most of the other stars. People used to joke they could read the newspaper by its light.

The star had planets of its own. There were five in all, and strangely enough all of them had rings, like Saturn. No one ever found any evidence of life on them, though.

At T minus sixteen years, Lucifer crossed the orbit of Pluto. Summer nights never really got dark after that—the intruder was vastly brighter than the full Moon. People had trouble sleeping. Lucifer's planets shouldered through the solar system. One came close to striking Neptune, zoomed past it in a swift arc, gripped that planet with its gravitational attraction, and slung it bodily away from the Sun and out of the solar system.

Over the next decade Lucifer drew so close it could be seen as a disk with the naked eye. People would hold up strips of exposed film to shade its light; through them the star looked the size of a pea. It was actually inside the orbits of Uranus and Pluto now. Uranus was on the far side of the Sun at the time, but Pluto, closer, had begun to move in Lucifer's direction. The orbits of all the planets were distorting. So too the Sun, which under its powerful gravitational pull was drifting toward the intruder. Each star attracted the other. Lucifer was falling at an ever-increasing rate into the Sun; the Sun was falling into Lucifer.

Two years before the cataclysm, Lucifer crossed the orbit of Saturn. One year later it passed within that of Jupiter, and by the time it had swung close by Mars (nearly vaporizing the planet) its velocity of approach had doubled. Scattered asteroids trailed after it like baby chicks hurrying behind their mother. Two months remained. It was summer then, and the heat was unbearable. Lucifer loomed overhead, the size of the Sun, each "night," which was as bright as day, and it filled the sky with an oppressive glare. One whole half of its red face was blotched and mottled. Vast radio storms continually drowned with static the few remaining TV broadcasts. Newspapers were full of nothing but doom. Over the course of the next month Lucifer swung from the nighttime to the daytime sky. Twin suns shone down upon a terrified world. Adults wept, and children ceased to play.

By T minus ten days, all semblance of normal activity had come to a halt. Temperatures were at the 150-degree mark. Lucifer was plummeting into the Sun at a speed of one hundred thousand miles per hour. It crossed the orbit of Mercury.

The last day dawned. Lucifer rose oval-shaped in the east. Minutes later the Sun rose, itself oval-shaped. Each star was radically distorted by the immense gravitational pull of the other. Monstrous storms crackled and flared upon their surfaces; vast prominences curled together and meshed. Over the next few hours the two stars moved visibly closer. Their distortions increased. Giant plumes of star-stuff rose bodily outward from each and licked about the other. People hid in cellars, stood in withered fields, prayed, swore, gazed upward, gazed into each other's eyes.

At half a million miles an hour, the two stars struck.

They splashed.

At the point of contact, a storm of superheated gas exploded outward. It shone with every color of the rainbow. Turbulent, immense, the spume boiled outward in an annihilating rush. In an instant the relatively cooler outer layers of the Sun were stripped away, and the hideous blaze of its inner core was revealed. A flare of overwhelming proportions flooded down upon the world: X rays, blue-white light. In a fraction of a second every forest and wooden structure on the sunlit hemisphere of the Earth burst into flame.

The debris of the mightiest explosion in the history of the solar system roiled outward. Moving soundlessly in the vacuum of space, it engulfed Mercury in its vaporizing fires, then Venus, then Earth. Slowing and cooling as it expanded, the cloud gradually grew more tenuous. At its heart—churning, chaotic, wildly unstable—finally stood revealed a new star, the amalgamation of what had once been two.

Many planets had been ejected in the collision, to wander

eternally through the near absolute zero of interstellar space. Others were gone without a trace—absorbed into one star or the other, or vaporized in the encounter. The remaining few were but cinders. And in the process every living creature had been wiped out, every city, every nation. Birds were gone, and trees and insects and grasses. The atmosphere had been stripped away, the oceans, even the outer reaches of the solid rocky mass of the Earth. The world had been sterilized. Not a trace of life remained on what had once been this good sweet Earth.

Grim fantasy. A shiver runs down my spine as I contemplate it. I am seated before my desk on a summer evening; the coming of night seems to have induced a certain ominous quality to my train of thoughts.

But reassurance is everywhere. A gentle breeze filters through the open window, bringing with it the scent of lilacs. Also in my garden are day lilies, honeysuckle, and a flowering apple tree, barely visible now in the gathering dark but strong companions all the same. A tiny rustle bespeaks some small beast—a chipmunk?—going about its business. Faintly to my ears comes the hoot of a distant owl.

I would not like to lose these things.

But the more you know, the more there is to fear. I know some things that complicate the apparently simple reassurance contained in the scene about me. I know how fragile it is. Forces of annihilation lurk everywhere, capable of wiping out every living being on the face of the Earth with the flick of a wrist. Presumably that chipmunk in the bushes does not recognize how precarious is its existence—and not just its own, private being, but that of the yard in which it lives, the acorns it harvests, and, indeed, the very Sun that brings it warmth. But I know.

Time for a stroll outside. It is fully dark now—no Moon tonight. Dimly sensed, a bat flutters overhead. I am fortunate to live in the country, and the air is crystalline. Overhead a sight greets me few city-dwellers have had the privilege to observe: the nighttime sky in all its glory. Across it stars are scattered like particles of gently glowing sand in a sandstorm.

Their number defies my imagination. Bright stars and dim, near ones and far, each one twinkling—happily, it seems to me now—and all arranged in constellations. But behind my pleasure lurks one small thought that renders all others problematic. I know those constellations are not eternal.

If I could gaze upward with a godlike vision in which thousands of years were compressed into seconds, I would see the constellations steadily altering. Each smoothly distorts as the ages pass. The constellations change because the stars of which they are composed are moving, and rapidly at that. And as for the pattern of this stellar motion, it is random. A given star in such and such a constellation might be speeding off to the left, the one just beside it roughly toward us, yet a third one downward. The motion of stars in the sky resembles nothing so much as that of mosquitoes in a swarm. This motion is superimposed upon the overall rotation of our galaxy and the still vaster expansion of the universe, but these are not the point. The point is the random swarming of the stars.

Our own sun is moving too, drifting through the swarm, the Earth and all the planets of the solar system carried with it on its journey: an immense journey, ponderous, deliberate—but also dangerous. For somewhere, untold light years distant and unknown to astronomy as yet, there is a star with our number marked upon it. It is on a collision course with the Sun.

A parking lot is filled with cars, all in rapid, frantic motion. Their drivers are acting without the slightest regard for safety, turning the steering wheels this way and that, stepping on the gas and slamming on the brakes, and all completely at random. Not only that, but every last one of them is blindfolded.

There are going to be some dented fenders soon.

That is the danger facing us. The peaceful scene about me is subject to the most deadly danger. How long do we have?

But hold on a moment! Don't concentrate on the future. Concentrate on the past, on all the thousands, even millions of years of history that have led up to this moment. The longer one waits, the greater the chance of collision, and if one had occurred at any point in the past nothing of what I see would have come into being. Had the Sun collided with a passing star in the epoch of the ancient Sumerians, none of us would have been born. The same would be true had the cataclysm occurred in the time of the dinosaurs. Our existence depends not simply on the avoidance of disaster this year or next, but throughout all of previous history.

Paleontologists tell us that the path leading to the emergence of *Homo sapiens* as a separate and distinct species was long and complex. It is difficult to pin down in the continuous progress of evolution the point at which our race appeared, but the emergence of language might serve as a convenient marker. This appears to have occurred something like one hundred thousand years ago. In human terms this is a vast stretch of time, but it is nothing compared to the age of the Earth as a whole, which is some 4.5 *billion* years. In cosmic terms our existence on this planet has been exceedingly brief, a mere flicker of an eyelash; and if a passing star had struck the Sun at any point in the preceding immense interval of time, humanity would never have come into being.

It's grim stuff. The more I stew over things, the more ways occur to nip humanity in the bud before it even got going. The passing star need not have actually struck the Sun in order to have wreaked havoc. In a near miss it could have dragged the Earth after it by its gravitational pull, detached us from orbit, and slung us off into the deadly cold of interstellar space, as illustrated in Figure 1. In a matter of months every living thing would have frozen to death. An even more distant passage would have left us in our orbit but would have distorted the orbit's form into an ellipse: As the Earth alternately approached and receded from the Sun it would have been alternately too hot and too cold to support life (Figure 2).

And finally, surrounding the solar system but at great distances from it is a vast cloud of comets, hundreds of billions of them. Even a far-distant passage of a wandering star would have been sufficient to deflect vast numbers of them toward us. The inner portions of the solar system would have be-

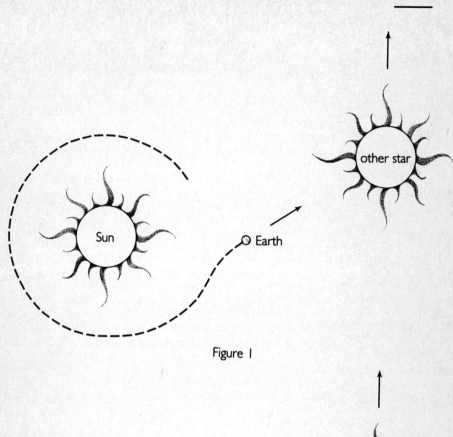

Figure 1

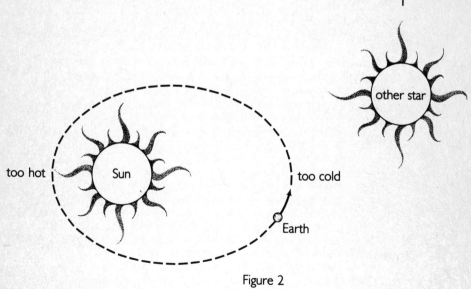

Figure 2

PROLOGUE

come flooded with comets, deadly projectiles flying randomly about. Had one of them struck our planet the energy released would have matched that of 100 million hydrogen bombs all going off at once, and great quantities of methane and ammonia, contained in comets, would have poisoned the air. Furthermore, the impinging comet would merely have been the first among many, and a world subject to such hammerblows at a steady rate would have been uninhabitable, a wasteland.

All these dangers lurk above me in the softly glowing stars. It's enough to give one pause. But that is not the point of my ruminations as I lie flat on my back, gazing upward into the abyss. The point is that they haven't happened. After all, I'm here, aren't I? In some way these disasters have not come to pass. Why not?

They have not come to pass because there is an element in the actual situation not covered in the analogy of cars speeding about a parking lot. This element is that parking lots are small but space is big. The stars are exceedingly far away, and as a consequence collisions with them are rare. How rare? Rare enough to keep us safe.

The remarkable thing about the arrangement of stars in space is the sparseness of their distribution. A scale model of the Earth and its place in the universe would make this clear. But it is difficult to construct such a model, and this because of the very sparseness of which I speak. Begin with a model in which the Earth is represented by a marble. In this model the Sun, the size of a medicine ball, sits three hundred yards away—but the star Alpha Centauri is a full forty-nine thousand miles away. And Alpha Centauri is not the most distant star. It is the closest star.

Try again. Make the model smaller; make the Earth a mote, a speck just barely big enough to see with the naked eye. Into that speck all seas, all mountains, plains, nations, and all the works of humanity have been jammed. Ten feet to one side now stands the Sun, an inch in diameter. On this scale Alpha Centauri is four hundred miles away.

A strange way to construct the universe. It appears to have

been designed by an extravagant, spendthrift hand. All that wasted space! On the other hand, in this very waste lies our safety. It is a precondition for our existence. Most remarkable of all is that the overall emptiness of the cosmos seems to have no other consequence in the astronomical realm. Had the stars been somewhat closer, astrophysics would not have been so very different. The fundamental physical processes occurring within stars, nebulas, and the like would have proceeded unchanged. The appearance of our galaxy as seen from some far-distant vantage point would have been the same. About the only difference would have been the view of the nighttime sky from the grass on which I lie, which would have been yet richer with stars. And oh, yes—one more small change: There would have been no me to do the viewing.

But if all these catastrophes will not come to pass, why bring them up? Why spend so much time worrying over disasters from which we are safe? *Because we are only just barely safe*—and that is the point of this book. Collision with a passing star is only the tip of an iceberg. Our existence, and that of every other life form in the universe, depends on a concatenation of circumstances, a network of interlocking conditions, each one of which must have held true in order for life to have come into being. The potential dangers that threaten us are so vast as to affect not just one person's existence, but that of life as a whole; and they arise not from circumstances such as war, pestilence, or famine, but from the very structure of the universe, from the nature of the laws of physics.

And yet here we are.

The burden of this book is not the dangers facing us. It is that none of those dangers have come to pass. But why have they not come to pass? The more one ponders this question the more mysterious it becomes. I believe that we are faced with a mystery—a great and profound mystery, and one of immense significance: the mystery of the habitability of the cosmos, of the fitness of the environment.

Over the past few decades much effort has been expended in searching for life elsewhere in the universe. The Viking

missions to Mars had this as one of their primary objectives:
Two unmanned vehicles landed upon that planet and conducted extensive biological tests there. Earth-based radio
telescopes have also searched long and hard for transmis-
sions beamed toward us by extraterrestrial civilizations. To
date all of these efforts have proved unsuccessful. Some scientists argue that these failures demonstrate life to be a rare
occurrence in the cosmos. We may be alone. An opposing
strong body of opinion, on the other hand, holds that it is
still too early to tell—we may not be alone. As for myself, I
wish to take no sides in this debate. It is not my concern. My
concern is to point out that all this immense, costly effort
would never have been expended were it not for a widespread conviction that it just might succeed. Scientists today
share the conviction that the cosmos is preeminently suited
to life.

They're right—the cosmos *is* suited to life. On the other
hand, by and large most scientists have taken this suitability
for granted. They have noted it and gotten on with the business of searching for that life. After all, at first glance it seems
unremarkable. But the central argument I wish to pursue in
this book is that a closer look would dispel this impression.
Indeed, I intend to argue precisely the opposite point of view:
that the habitability of the universe is an utterly astonishing
thing. In what follows I hope to persuade the reader that,
whether life is common or rare, the fact that it has arisen at
all on the cosmic scene is profoundly mysterious. The deeper
one looks, the more surprising it becomes that we exist.

My method of persuasion will be to describe a variety of
ways in which the universe might have been unsuitable for
life—but is not. My method will be to detail what can only
seem to be an astonishing sequence of stupendous and unlikely accidents that paved the way for life's emergence.
There is a list of coincidences, all of them essential to our
existence. That we have not collided with a passing star is
only the beginning of the story.

It is one thing to assemble a list of conditions required for
life to exist, and quite another to decide that the list adds up
to an enigma. After all, it would always be possible to come

up with the conditions required for me as an individual to exist, and it is clear the list would make dull reading: I must have been born, must have been cared for during infancy, must have partaken of food regularly, and so forth. Nothing here is cause for remark. So why get excited about the set of conditions required for life in general to exist?

The answer is that these conditions turn out to be of an entirely different sort. Each and every one of them is surprising, and they are surprising because they involve striking and remarkable coincidences. The best analogy I know of to the condition of life in the universe has been given in an article on the subject by the Canadian philosopher John Leslie. Leslie's analogy concerns a man sentenced to be shot at sunrise. Early in the morning he is dragged before the firing squad. He stands before it blindfolded, the commander gives the order to shoot . . . and by some extraordinary and unprecedented chain of coincidences, every last one of the rifles in the squad misfires.

Our failure to collide with a passing star is the first of all those defective rifles in the firing squad. Part One of this book will list the others.*

This list was not invented by me. It has been amassed over the decades as the result of the patient efforts of scores of scientists. These efforts have intensified in recent years, in my opinion primarily due to the codification by the physicist Brandon Carter of the seminally important Anthropic Principle. This principle is discussed in Chapter 2. Its name derives from the Greek *anthropos*, meaning "man"; anthropic studies are those aimed at elucidating the conditions required for mankind to arise within the universe.

Some of the work described in the pages to follow was explicitly focused on the question of life in the universe. Examples are the inquiry into the red giant stars described in Chapter 1, that into the properties of water (Chapter 4), and that into the dimensionality of space (Chapter 8). Other scientists, on the other hand, could not have been less interested in the question. A prime example is the subject of high-energy

*The list is summarized in the Appendix.

physics and its bearing on the very early history of the universe, discussed in Chapters 10 and 11. The authors of these ideas were interested in cosmology and the nature of elementary particles, not life. No matter: Whether they intended to or not, they ended up adding crucially important elements to the picture.

But among all the scientists who have thought about the subject, only a very few have taken the last crucial step of emphasizing the essential mystery of life's existence. The first was Lawrence J. Henderson, professor of biological chemistry at Harvard around the turn of the century, who in 1913 published a profoundly significant book entitled *The Fitness of the Environment*. Although Henderson lived long before the formulation of the Anthropic Principle and the contemporary surge of interest in the subject, his work is strikingly modern in many ways. And at present George Wald, also at Harvard and a Nobel laureate in physiology and medicine, is in the process of carrying Henderson's ideas further. Time after time Wald has argued forcefully and persuasively for the existence of a major enigma confronting us.

Most other people interested in the Anthropic Principle, however, have avoided this step. They have contented themselves with the task of assembling the list of conditions required for our existence and letting it go at that. Science proceeds by identifying problems and then solving them. But before you do the second you have to do the first. You must decide there really is a problem to be dealt with. Fortunate indeed is the scientist who has identified a new conundrum, a problem so impressive that it attracts immediate and widespread interest. The problem is likely to be solved in short order and the scientist's reputation is assured. Fortunate too is the investigator whose new idea attracts immediate and widespread opposition—at least the idea is being taken seriously. But in my own experience whenever the problem of the fitness of the environment is broached, most scientists respond neither with interest nor with opposition. They respond with a smile. They respond with a shrug.

That is the most damning response of all. The problem has been rejected, and the rejection is instinctive. It is gut-level, unmediated by rational thought, and it takes place prior to the point at which any serious consideration is possible. Before anyone has had time to think, the subject has been dismissed.

And perhaps now a bit of personal history is in order. When I first became attracted to Carter's Anthropic Principle, I regarded it as a matter of academic interest only. I figured it would be amusing to know the conditions required for life to arise in the universe—amusing and probably instructive, but hardly of great significance. At the time I was entirely unaware of Henderson's work on the fitness of the environment, and of Wald's long series of articles on the subject. But as I read the works of other scientists, I set myself the task of summarizing their conclusions in the form of a list, an actual piece of paper sitting before me on the desk. Initially that list occupied a scrap torn from a notepad. I kept reading; soon it occupied a more official 8½-by-11 sheet, then several of them. The list kept getting longer . . . but that was not the point. The point was its strangeness. So many coincidences! The more I read, the more I became convinced that such "coincidences" could hardly have happened by chance.

But as this conviction grew, something else grew as well. Even now it is difficult to express this "something" in words. It was an intense revulsion, and at times it was almost physical in nature. I would positively squirm with discomfort. The very thought that the fitness of the cosmos for life might be a mystery requiring solution struck me as ludicrous, absurd. I found it difficult to entertain the notion without grimacing in disgust, and well-nigh impossible to mention it to friends without apology. To admit to fellow scientists that I was interested in the problem felt like admitting to some shameful personal inadequacy. Nor has this reaction faded over the years: I have had to struggle against it incessantly during the writing of this book. I am sure that the same reaction is at work within every other scientist, and that it is this which accounts for the widespread indifference accorded the idea at present. And more than that: I now believe that what appears as indifference in fact masks an intense antagonism.

It was not for some time that I was able to place my finger

on the source of my discomfort. It arises, I understand now, because the contention that we owe our existence to a stupendous series of coincidences strikes a responsive chord. That contention is far too close for comfort to notions such as:

We are the center of the universe.

God loves mankind more than all other creatures.

The cosmos is watching over us.

The universe has a plan; we are essential to that plan.

Anthropocentrism is the attitude lying behind these notions. Each reader will have his or her own reaction to them. It is important to make clear that I myself am not religious, and (a separate issue) that I find anthropocentrism utterly distasteful. Furthermore, I can testify from personal experience that I am not alone among scientists in this matter. The roots of the widespread distaste for the concept of a universe suited to life run deep, and they are intimately connected with the very nature of the scientific enterprise. The scientific revolution was in large part a reaction against anthropocentrism, an attitude expressed in the old geocentric cosmology, which placed the Earth at the center of the universe. In a similar reaction, the Darwinian revolution brought forth the theory of evolution, a theory that dethroned humanity from its presumed central position in the overall scheme of life. Each of these advances was furiously resisted at the time—and in the case of the theory of evolution, the resistance has continued to this day, as evidenced by the tireless efforts to include so-called creation science in the curricula of our public schools.

No scientist worth his salt would tolerate for an instant a return to the prescientific mentality enshrined in notions such as creationism. I applaud this attitude, and I share it. On the other hand, for reasons given in Chapters 9 and 12, I believe that there is no relationship between the anthropocentric mentality and the central thesis of this book. The widespread antagonism accorded the thesis is based on a misapprehension—anthropism is not anthropocentrism.

Of course I would never claim the two do not *appear* the same. They do indeed "smell" alike. And as a consequence, the idea that the universe is radically suited to life has an uphill battle to fight if it is ever to gain acceptance.

*　　*　　*

Ask a new question and you will learn new things. But like all other new things, right now the Anthropic Principle's implications are vague and undefined. I cannot clearly see where all this will end. Furthermore, no single person could possibly establish a whole new view of physical reality single-handed. At best I can hope only to render plausible the contention that, in some strange and at present mysterious fashion, our universe is fundamentally a universe of life—a universe that takes life seriously, if you will. Only when enough people begin to take the idea seriously will further evidence leap forward as if spontaneously.

And make no mistake about it: If that idea turns out to be correct, it is no exaggeration to say that a major revolution in thought is in the offing. Whatever the explanation turns out to be for that massive series of coincidences whereby life arose in the universe, it is not going to be simple. Each and every one of them flows from the laws of nature, and it is to these principles themselves that our thinking must turn. Life obeys the laws of physics—this much is a truism. What is new and incomprehensible here is that in some extraordinary way the reverse seems also to be true—that the laws of physics conform themselves to life.

How could this possibly have come to pass? Part Two of this book takes a first cut at the question. As we survey all the evidence, the thought insistently arises that some supernatural agency—or, rather, Agency—must be involved. Is it possible that suddenly, without intending to, we have stumbled upon scientific proof of the existence of a Supreme Being? Was it God who stepped in and so providentially crafted the cosmos for our benefit? Do we not see in its harmony, a harmony so perfectly fitted to our needs, evidence of what one religious writer has called "a preserving, a continuing, an intending mind; a Wisdom, Power and Goodness far exceeding the limits of our thoughts?"

A heady prospect. Unfortunately I believe it to be illusory. As I claim mankind is not the center of the universe, as I claim anthropism to be different from anthropocentrism, so too I believe that the discoveries of science are not capable of proving God's existence—not now, not ever. And more than that: I also believe that reference to God will never suffice to explain a single one of these discoveries. God is not an explanation.

What then? If we cannot explain in a religious fashion those remarkable coincidences whereby life arose in the cosmos, how are we to understand them? Chapters 13 and 14 explore a remarkable possibility suggested by the great revolution in thought that is quantum mechanics. This theory was developed in the first three decades of the century in an effort to understand atomic structure; the problem of life in the cosmos could not have been more distant from the minds of its inventors. Nevertheless, insights gained from the theory may have great relevance for the problem. It may be that the explanation for nature's extraordinary fitness for life must be sought not in the realm of religion, not even in any purely scientific realm, but in the realm of existence itself. Metaphysics, the study of existence and of the ultimate nature of reality, is usually considered part of philosophy. But quantum mechanics too has something to say about the subject.

The insight suggested by this theory—and I emphasize the word *suggested* here—is that in the fitness of the environment we are witnessing the effects of a gigantic symbiosis at work in the universe. Symbiosis, the mutual interdependency of two organisms, is widely known in biology, but the symbiosis envisaged here is different. The first partner in this new relationship is not an organism at all, but rather an inanimate structure: the physical universe as a whole. As for the second, it is alive but it is not any single organism. It is all organisms—life itself.

And between the two there is a union. There is a great metaphysical dance by which each supports the other. How did it come to pass that against all odds the cosmos succeeded in bringing forth life? It had to. It had to in order to exist.

Well . . . big words. Pretty tough talk from so small a fellow: one tiny individual flat on his back under this great impersonal sky. It strikes me I make a foolish figure right now, declaiming so confidently the latest poop on the structure of all things—stars, cosmos, and yes, even the nature of existence itself. While I've been lying here some mosquito has found me out. It evades my every slap. Obnoxious beast— who gave *it* permission to exist?

It's late. Time to be moving. I get to my feet and stretch. But before heading indoors I pause a moment and look about.

There is not a breath of wind. There is not the slightest sound. That chipmunk in the bushes seems to have closed up operations for the night. Overhead, the stars are strewn across a darkness, a blackness so profound that for a moment, for the barest flicker of an instant, I can almost sense their inconceivable distance. In a sudden, exalting burst of vertigo I fancy what it would be like to fly, to fall up and into that ocean. And in my imagination I am falling now, falling slowly, falling endlessly, tumbling gently through the stars in the great and perfect isolation of the night.

PART ONE:

LIFE

I

The Red Giants

There was a game I used to play as a child that now, for some reason, comes back to me. It was to identify each and every substance in my view. Not so easy as it sounds, for the more you look the more you see.

I am moved to try it again. Let's see . . . there is the piece of paper on which I write these lines, and the ballpoint pen with which I write them. Without moving from my chair I can spot the rug upon the floor (wool there), the wood of that floor, and the glass of windows. Simple so far—but there's more. The ballpoint pen turns out on close examination to be built of plastic wrapped about a metallic core, not to mention the ink itself. Each element of the complex pattern on the rug bespeaks a different dye. My clothing—I had not noticed it before—is composed of wool (sweater), cotton (shirt and pants), plastic (buttons), animal skin (shoes, belt), and metal (zipper). And as for the TV set in the corner, it grows more daunting the longer I think about it: a complex structure of metal, plastic, glass, and Lord knows what else, contained in the electronic maze of its interior.

An incomplete list, of course. I ought to include all the trees, leaves, bushes, stones, birds, and insects in the backyard; all the streets, sidewalks, cars, fire hydrants, and billboards fading off into the distance; all the roots, grubs, grains of dirt beneath ground level. The task is hopeless.

That's right. The task *is* hopeless, and it is hopeless because the world is a complicated place. The more one studies it the more layers upon layers of pattern, structure, and detail emerge. As a matter of fact, this organism seated before a piece of paper right now and scribbling vigorously away is by far the most complex structure in sight. Imagine the task of cataloging all the tendons, ligaments, veins, nerves, bones, organs, and glands of which I am constructed!

There is a point to this game—two points, in fact. The first is that this complexity is utterly essential to our existence. The second is that it would not exist if it weren't for stars with names like Pollux, Capella, Arcturus, and Aldebaran.

The absence of collisions with a passing star, an absence made possible by the vastness of space, is one of the preconditions for life. Complexity is another. All of us maintain a certain level of richness and variety throughout the day in order to keep boredom at arm's length, but that is not my point here. My point is that a truly simple universe could not possibly contain life. "Simplify!" exhorted Henry David Thoreau from the shores of Walden Pond. But not too much.

Even the simplest of organisms, a mite, an amoeba, or a bacterium, turns out to possess an enormous wealth of detail when examined carefully. There are no uncomplicated living beings. Biologists have always had difficulty in defining just what they mean by life, but all would agree with Greenstein's quick and handy rule of thumb: If it's simple, it's dead. As a matter of fact this wealth extends all the way down to the atomic realm. A rich composition in terms of the chemical elements is a fundamental hallmark of life. Many different

elements are employed in one fashion or another in the biochemical workings of an organism, and not just of some organisms, but of every organism, and in every cell. There is
not a single life form known to science that makes do with
only two, or even twenty, of them. This tendency toward
complexity of composition stems from the enormous richness
of chemistry, the magnificently varied interactions of the elements. Every atom acts for all the world as if it were fes-
tooned with a variety of grappling devices—hooks, clamps,
pincers, bits of Velcro, spots of tar: a different set for each
element, each device operating differently, and each specifically and uniquely tailored to grapple in differing ways with
differing mates. Chemists have cataloged tens of millions of
different chemical reactions; theoretically speaking there is
no limit whatsoever on their number, or on the variety of
molecules so formed. And life depends for its workings upon
the infinitely varied possibilities provided by this richness.

Among the chemical elements hydrogen is the simplest,
helium the second simplest. It is instructive to ask whether
life could make do with just these two. The answer is a re-
sounding and unqualified no. Why? In the first place because
these two elements are gases. Except under highly unusual
circumstances of enormous pressure or extreme cold, they do
not form solids or liquids at all. In such a gaseous world I
would find no pen and paper, no clothes, wood, or walls, no
ground beneath my feet—not to mention the feet themselves.
Nor would there be any "me" to do the finding. The world
would consist only of a diffuse and tenuous mixture not so
very different from air.

But solids and liquids are crucial to life and its workings.
Among all the myriad organisms known to biology, not one is
made of air. They may live in air—birds, mosquitoes—and
they may require it for their functioning—we breathe—but
they are not composed of it. This is doubly striking in that
the three chemical elements most universally employed in
the biochemical machinery of life, carbon, nitrogen, and oxygen, are abundantly present in the atmosphere; indeed, they
are its primary constituents. It is the restraining, ordering
nature of solids that accounts for their universal presence

within living beings. Organisms need to have a shape. They require a fixed and invariable structure, and cannot diffuse about like a puff of wind. Evolution has provided for this need by employing solids in their construction. As for liquids, they too are mobile: Life appears to find it quite impossible to make use of the complete anarchy a purely liquid construction would allow. Here too, order is required.

The second reason a hydrogen-helium cosmos could not support life appears at the atomic level amid the richness of chemical interactions upon which life depends. Hydrogen and helium do not have complicated chemistries at all. Their chemical interactions are deadly dull. Helium in fact has no chemistry. It is inert, and if placed in contact with other atoms it will do exactly and precisely nothing. Hydrogen in conjunction with helium also undergoes no reactions. Hydrogen in conjunction with more of its kind does react: Two atoms combine to form one hydrogen molecule . . . but this, in turn, is gaseous and inert, and incapable of any further chemical transformations.

So if the task of cataloging the world before me is far beyond my reach, that of cataloging a hypothetical hydrogen-helium universe is well-nigh trivial. It would consist of a gas with only the most rudimentary chemical properties. A dead gas.

Pollux is a brilliant star, one of the pair making up the constellation of Gemini, the twins. It is overhead in winter and early spring. Capella sits not far from Pollux, also visible in winter, embedded in the Milky Way. Arcturus is up in summer; find it by tracing the arc of the handle of the Big Dipper across the sky till it leads to that bright star—"Arc to Arcturus," people say. Aldebaran is the brilliant eye of Taurus, the bull.

These stars as they shine so brightly are in the process of creating complexity—of creating complex elements. Ages ago there occurred the Big Bang in which the entire universe as we know it began. Rushing outward was a superheated cloud of hydrogen and helium. Why hydrogen? Because only this, the simplest of elements, was capable of withstanding that blaze.

Just as the heat of an ordinary fire will reduce a composite structure such as a chest of drawers to a simpler one such as ash, so the heat of the Big Bang was sufficient to reduce any complex atoms that might have existed back then into the simplest. And why helium? Because nuclear reactions occurred in the bang that synthesized it from the hydrogen.

So that is how the universe began: hydrogen and helium.

Immediately after its creation the universe was therefore unfit for life, utterly so. Something must have happened to transform it into the more suitable abode we find today. This "something" was the nuclear transformation of the elements. Hydrogen was amalgamated with more hydrogen to yield helium, then helium with helium to yield carbon, and so on up the periodic table. The process took place within stars, some of which exploded, others of which more sedately expelled the elements so created into the vastness of interstellar space. In either event, these elements were then amalgamated into new stars and planets . . . and so it was that complex atoms synthesized in an ancient sun eventually wound up in a lukewarm tidepool beside a prehistoric sea. The stage was set for the emergence of life.

It took a long time. As a matter of fact, it is still going on; bit by bit, as the millennia roll by, the composition of the universe is systematically altering to include a greater and greater proportion of the heavy elements. It is a process the ancient alchemists forever dreamed of but never achieved: transmutation.

They never achieved it because to change one element into another requires very high temperatures. How high? Ten million degrees will do. There is a profound difference between the *nuclear* reactions responsible for this transmutation and the *chemical* reactions the elements undergo once they are so created. At relatively low temperatures—that of a flame, say, or a blast furnace—only chemical transformations can occur. But heat things up to millions of degrees and an entirely new process will commence.

Where is matter heated so greatly? Only in stars. The heart of the Sun is maintained at such terrible temperatures, and as a consequence nuclear reactions are proceeding within it at this very moment. The Sun is in the process of transmuting hydrogen into helium. So are many other stars in the sky.

But these make only helium. There is a second step, the transformation of helium into carbon. However this, it turns out, cannot occur in the Sun, nor in stars like it. A different sort of beast is required to yield this reaction. Capella, Pollux, Arcturus, and Aldebaran are examples of this beast.

Although it is not obvious from their appearance in the nighttime sky, these stars are quite different in structure from the Sun. In the first place, they are unusually brilliant. In the second, they are deep red in color, and they are very large—up to fifty times larger than more ordinary stars; that is the difference between a person and a good-sized apartment building. Hence their name: They are the red giant stars. And as a consequence of their unusual structure they are capable of doing what the Sun cannot, of synthesizing carbon. This must also have occurred ages past, shortly after the cosmos began: a slow, grand evolution of the stuff of the world, the creation of elements required for life by an earlier generation of red giant stars.

Clearly it happened—for here we sit, endowed with all the complexity one could desire. But how did it happen? For it turns out there is a problem. There is a roadblock standing in the way. This roadblock prevents the creation of carbon, and indeed of every element beyond helium in the periodic table. It stands in the way of the formation of all those complex, reactive elements upon which life so crucially depends. The red giant stars manage to find their way around that roadblock, but for years no one was able to understand how. And even now, when the answer is well in hand, the method they employ seems astonishing. Deep inside them an extraordinary coincidence is at work. Three quite different structures work together in a wildly improbable fashion. There is a harmony present. It resembles nothing so much as the other, the musical harmony that delights the ear. The best way I know of to think about it is in terms of the tuning of musical instruments.

*   *   *

Nuclear reactions take place between the nuclei* of atoms. These nuclei are composed of two different subatomic particles, the proton ⓟ and the neutron ⓝ. The red giant star contains helium, whose nucleus is made up of two protons and two neutrons (Figure 3), and within it these nuclei react together to produce carbon, which contains six of each particle (Figure 4).

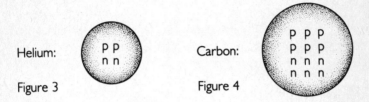

Helium: p p / n n   Figure 3   Carbon: p p p / p p p / n n n / n n n   Figure 4

A glance at these diagrams is enough to reveal the nature of this reaction. A carbon nucleus is made by putting three helium nuclei together. One way to do this is simply to take three heliums and jam them together by brute force: They will combine to yield a carbon. Of course that is not the way it goes in the red giants. They are gaseous, and in a gas every particle—every helium nucleus—is flying randomly about. You simply wait until, by chance, they collide.

You wait till *three* collide, and all at the very same instant. But that is a colossally unlikely event. While ordinary collisions are common, the chance of such a triple encounter is very low. It is, in fact, too low by far to allow the production of carbon in this way.

How, then, is it produced? The only other possibility is a two-stage process. First *two* heliums collide and produce some intermediate nucleus. That is not such an unlikely event, and it will happen rapidly. Next this intermediate nucleus collides with another helium to make the carbon.

What is that intermediate nucleus? It comes from two heliums, so it must be made of four protons and four neutrons (Figure 5).

---

*Definitions of technical terms can be found in the Glossary.

THE RED GIANTS

Figure 5

That is beryllium, an element whose weight is intermediate between those of helium and carbon. But it is not the usual form of beryllium. Normal beryllium has one more neutron, as in Figure 6.

Beryllium:

Figure 6

The stuff that is formed in red giants is an abnormal form, known as an isotope.

The problem is that this particular isotope of beryllium is unstable. It flies apart the instant it is produced. It is, in fact, the most wildly unstable isotope of beryllium known. That element exists in four forms. The first, the normal form, is stable and survives forever. The second survives a full 3 million years before disintegrating, and the third lasts two months. Had the nuclear reactions in red giant stars yielded any of these three forms, there would have been no problem. But instead they yield the fourth—and this is a form that bursts apart in a mere 0.000000000000001 of a second.

Life in the universe depends upon the creation of heavy elements. These follow upon the creation of carbon, and this, in turn, comes from an isotope of beryllium so fantastically fragile that it can hardly be said to exist. Upon so desperately frail a house of cards does our existence depend.

That is the roadblock. The isotope is far too unstable to allow the formation of carbon.

*   *   *

How do the red giants get around it? For years no one knew. The first step toward an understanding was taken in 1953 by the Cornell University astrophysicist Edwin Salpeter. Salpeter invoked the notion of a resonance between the helium and the beryllium nuclei. What is resonance? Its nature can be illustrated with the help of an example from music.

Take a tuning fork tuned to some note, A perhaps. I will take another, also tuned to A. Stand beside me and give yours a whack. As you do so I will *not* strike my own—I'll just stand there. Now halt the ringing of your fork, by stuffing it in a pocket, say. A faint, clear note still resounds. It is coming from my hand. Your act of striking your fork has set mine to ringing.

That's resonance. Now do another experiment. You choose an A fork but I'll choose B. This time it will not work. My fork will not be set ringing when you strike yours. They have to match—they have to resonate.

Resonance is not confined only to music. I employ it every time I visit the playground. The swing in my neighborhood playground has a certain natural rate of swaying to and fro: once a second, let's say. So when I push a child upon it, I push once a second. I adjust my rate to match the swing's—I work in resonance with it. If I don't, we will never get anywhere, the child and I. I will push and shove, sometimes working with it, sometimes not, and no great motion will result.

In this example the resonance is between two rates: that of the swing on the one hand, that of pushing on the other. In the previous example it was between musical pitches (which turn out to be the same thing, though the fact is not significant here). In nuclear physics, the resonance is between two energies.

Just as a tuning fork resounds with a pure tone, just as a swing oscillates at a definite rate, so every nucleus possesses a unique energy. You cannot force it to have a little more or a little less: That energy is fixed by the fundamental laws of physics. The helium nucleus has such and such a value; that fragile, wildly unstable isotope of beryllium another. But what Salpeter realized was that those two energies resonated.

Only if tuning forks resonate will striking one set the other singing. Only by pushing in resonance with a swing will I set it moving efficiently. Similarly, Salpeter realized, because it was resonant with helium, beryllium would be produced from helium at an enormously enhanced rate. The thought occurred to him that under these circumstances it might not matter that beryllium decayed so fast. Perhaps it would be re-created just as rapidly as it flew apart. Perhaps the stuff would be present after all, available to participate in the next stage of the nuclear reactions, the stage that made the carbon.

That was Salpeter's proposal. Unfortunately it did not work. Detailed calculations were performed, and they showed that even with his new twist the roadblock remained. The expected concentration of beryllium in red giant stars persisted in remaining intolerably small—about one part per billion.

It was the British astrophysicist Fred Hoyle, then at Cambridge University, who supplied the answer. Hoyle took Salpeter's notion further and invoked yet another resonance, this time in the second stage of the nuclear reactions. He asked whether the energy of the reacting beryllium and helium nuclei matched that of their final product, carbon. If it did, the production of carbon from beryllium would likewise be enhanced, and the roadblock surmounted.

Hoyle checked it out. The structures of these nuclei had already been studied. Tables of their energies were available. Poring over them, he found the energies did not match. No good.

He pushed it further. He knew that nuclei had not one but many resonances. Here too, the musical analogy holds. An A tuning fork will resonate not just with A, but with other notes—with the overtone series of A. There is a simple rule, familiar to musicians, giving the set of all notes in resonance with a particular one. Nuclei are more complex and no such rule is known, but the principle is the same; one can explore them experimentally, probing their structures with a nuclear accelerator, and work out their various energy levels, the equivalent of their overtone series. This too had already been performed, and the results were available in still more extensive tabulations. Hoyle pored over these. Failure. Nothing matched. No resonance . . . and no carbon, no life.

So he made them match.

A different person would have given up, proclaimed the idea no good, and gone on to something else. Fred Hoyle reared up on his hind legs and declared the experimenters had not been careful enough. He insisted they had missed one of carbon's resonances. He sat down, and with paper and pencil calculated exactly where that resonance had to be, by insisting it match with the known energies of the helium and beryllium nuclei. And then he went off to tell everyone all about it.

In the entire history of science this is the only example I know of in which something was actually predicted on the basis of such reasoning. Hoyle spent the next year on sabbatical leave, visiting a nuclear physics laboratory at Caltech. He arrived big and brash, full of beans, and informed people of the situation. Possibly because of his insistence, possibly because he was a personal friend of the lab's director, several of the physicists there took up his notion. Finding resonances is usually a difficult task, something like looking for needles in a haystack. These physicists' task was easier, though, for they knew where to look. Hoyle had told them.

They looked. And they found it.

There are three quite separate structures in this story—helium, beryllium, and carbon—and two quite separate resonances. It is hard to see why those nuclei should work together so smoothly. Others do not. Other nuclear reactions do not proceed by such a remarkable chain of lucky breaks. It's not so unusual to find resonances in nuclear reactions, but such a double matching is unique.

All in all it is a remarkable coincidence. In assessing the magnitude of this coincidence, its degree of probability, the musical analogy fails, for there are after all only a limited number of notes. It is not like reaching randomly into a drawer stuffed with tuning forks and by good fortune pulling out three A's. Rather it is like discovering deep and complex resonances between a car, a bicycle, and a truck. Why should such disparate structures mesh together so perfectly? Upon this wildly unlikely coincidence our existence, and that of every life form in the universe, depends.

2

The Anthropic Principle

It strikes me I'd like to know the explanation for that coincidence of resonances between helium, beryllium, and carbon. Is it an accident?

The answer to that question depends on one's point of view. On the one hand, there is nothing supernatural going on. In this regard the analogy of the car, the bicycle, and the truck is a good one: We know how they're built. If by some chance these three vehicles were to vibrate in resonance with one another, engineers poring over the blueprints would be able to come up with an explanation. Similarly, the principles of nuclear physics are relatively well in hand. If we but knew the fine details, we would have been able to predict the resonances from a knowledge of the structure of those nuclei.

All perfectly true—but it is not enough. Suppose I were to shoot an arrow at a target by squeezing my eyes tightly shut, spinning about madly on one foot, and then loosing the bow completely at random . . . only to find, upon opening my eyes for a look, the arrow lodged exactly in the bull's-eye. The fact is irrelevant that in its flight it obeyed precisely the laws of ballistics, that even as I spun the motion of my body followed known physical principles. The situation would still be a matter for some surprise.

Imagine similarly an immense, a gigantic stack of pages, each one of them covered with numbers, equations, and formulae, blanketing a desk. You are looking at a complete mathematical theory of the structure of helium, beryllium, and carbon. Rummage through the heap. At last you find, buried toward the bottom of some obscure page, a formula. A formula stating that they resonate. *Stating that you are allowed to be.*

It is difficult to be content with such an account. No matter how mathematically correct it would be, still it would not satisfy. For a coincidence of such immense significance to humanity we want a deeper account. Where to find it?

Here's a thought: One "account" is that we exist.

The proposal is that we know those nuclei match so perfectly because we know we exist. At first blush the proposal may seem to have things backward. After all, in truth the chain of causation runs the other way: from the existence of the resonances to the existence of us. But no one is claiming the nuclei match because we're here. The claim is that *we know* they match because we're here. And this is just the argument Hoyle used.

It was in 1974 that the British physicist Brandon Carter, now at the Observatory of Paris, coined the term Anthropic Principle to describe the curious, backward sort of reasoning Hoyle had employed. The Anthropic Principle is the statement that *if some feature of the natural world is required for our existence, then it must indeed be the case.* Such a statement cannot possibly be doubted. It is logically true: true as only a tautology can be. If a child's shoes are tied, someone must have tied them. If children are born from mothers, that child was born from a mother. And if three submicroscopic particles in the hearts of distant stars must have resonated together in order for us to exist—why then, they did resonate. Our existence implies that of those resonances.

There's more, of course. Our existence also demonstrates the stars to be exceedingly far away. Imagine an alternate universe in which the stars were not so very distant. How distant? Choose an intermediate situation in which their packing was sufficiently sparse that direct collisions between stars hardly ever occurred, but sufficiently dense that more distant passages regularly removed planets from orbit about their stars. In many ways that imaginary cosmos is not so very different from our own, containing as it does stars, planets, and galaxies. From a distance all would seem normal. Up close, though, a problem would become evident: The Earth, tugged from its orbit by a passing star, would be wandering freely through interstellar space. Deprived of its warming Sun, the Earth would be cold—bitterly cold, close to 460 degrees below zero Fahrenheit. As would be every other planet.

A question now. In that imaginary cosmos, would there be anyone around to know how cold is was? Are we allowed to add to the picture another element, that of people huddling together for warmth in the eternal blackness?

No, we are not. Those people could not survive. More than that: They never would have been born in the first place.

There could be no one present to realize the stars were close. Such a proximity of stars is a thing that cannot be experienced. In the absence of knowers, it cannot be known. And a different imaginary cosmos in which the three nuclei did not match so precisely would also be one that could never be experienced. It would forever remain an undiscovered country, and the structure of those nuclei could in very principle not be elucidated. Their failure to resonate would never be discovered. In a very real sense an uninhabitable cosmos is an invisible cosmos—one not capable of being seen.

The only things that can be known are those compatible with the existence of knowers. That is the Anthropic Principle in its purest form.

The Anthropic Principle is a remarkable device. It eschews the normal methods of science as they have been practiced for centuries, and instead elevates humanity's existence to the status of a principle of understanding. It operates in the realm of alternate realities, asking what the universe would have been like had some feature of reality been different, and whether life would have been possible in such a circumstance. If the answer is no, we can be sure that feature could not have been other than it is.

It involves a pretty trick of logic. Raise up your hand and look at it; you are observing yourself. But you are part of the universe—as you look, the very cosmos is observing itself. Now turn and glance at a tree. The universe is still observing itself. Study the distances to the stars and the structure of atomic nuclei; still, one piece of the universe is studying another. Through our existence the cosmos has become self-reflective. But a cosmos devoid of life is unable to perform this remarkable trick.

One of the functions of those resonances is to ensure that their existence be known.

Here is a railroad track. On it is a train. Ages past that train set forth upon a great journey, a voyage of inconceivable magnitude; its track twisting and turning, winding through forests, valleys, and mountain passes, skirting great rivers, prairies, mountain lakes. So vast is that journey, so far-distant the destination of the train, that babies are born upon it, grow old, and die. Generation after generation has passed in this way.

Somewhere beside the railroad track a great red X has been painted on the ground—no other such markers scattered along its length, but one and only one. The train has no windows. One day a curious soul decides he'd like to take a look outside. He drills a hole in the wall. Bending down to peer through, he spies something interesting. It is an X painted on the ground. Amazing coincidence to have decided to look at just the right moment!

The track is the winding path of cosmic history, beginning with the creation of the universe and extending on into the unseen future. The train upon it is the moving point representing the present. That curious soul represents us; the drill, our modern understanding of the physical world. The coincidence is a remarkable agreement between two numbers, numbers that by rights have no business agreeing with one another. For decades no one understood that agreement. It led to proposals, counterproposals, speculations bordering on the mystical. The resolution was finally achieved by the Princeton University physicist Robert Dicke (pronounced *Dickey*). He did it using the Anthropic Principle, and even today his work constitutes one of the prettiest uses of that principle I know of.

Dicke's mystery concerned two so-called cosmic numbers, so termed because they express certain fundamental properties of the cosmos. The first of these numbers refers to time. It asks how old things are. The house in which I live is fairly old, but not nearly so much so as the material out of which it is built. Before that house existed the wood in its walls stood as tall trees, the iron in its nails lay hidden as ore. And what about the ground upon which it sits, the dirt, the stones— how old are they? We are not accustomed to thinking about such things, for in human terms the world seems eternal. But it is not eternal. Ages past what is now my land was flooded by an ocean, before that it sat atop a mountain . . . and before that it did not even exist. The Earth was created; by current reckoning, about 4.5 billion years ago.

On the other hand it is only this planet that came into being at that time, or more precisely the solar system as a whole. The *matter*, the atoms and molecules out of which we would ultimately be constructed—this matter existed before the Earth did, freely floating in space as giant interstellar clouds. The question is whether matter is eternal.

One of the most profound discoveries of twentieth-century science is that it is not. Matter has an age. Substance, material, the pure primordial stuff of the world—it came into being in the Big Bang, and this occurred at a very definite moment in time. To the best of our knowledge, before that moment nothing existed; after it, everything did. We do not know very accurately how long ago that extraordinary mystery of creation was enacted. Seventeen billion years is a ballpark figure, although it is very uncertain. Somewhere in there lies a number of the greatest significance, the lifespan of the fabric of the universe.

The first cosmic number measures this lifespan. It is not

the age of the universe in years, however. After all, the year is merely the length of time required for our particular planet to orbit our particular sun, and there is nothing very important about it in the overall scheme of things. To get a number whose significance is more profound we need to work in truly fundamental units. An appropriate one is known, cheerfully enough, as the jiffy. The jiffy is the time required to traverse a fundamental building block of the cosmos, a proton, while traveling at the most fundamental velocity known, the speed of light. Light is fast and particles are small; the jiffy works out to some 0.0000000000000000000000009 of a second. The first cosmic number is *the age of the universe expressed in jiffies.*

That number is 66,000,000,000,000,000,000,000,000,000, 000,000,000,000. Write it down on a piece of paper. Carry it with you always. Tack it up on bulletin boards; scrawl the number as graffiti in bathroom stalls. That pure, abstract mathematical object comes as close to being holy as science will allow. The ancestry of matter is revealed within it.

The second cosmic number refers to force.

This number compares the strengths of two of the fundamental forces of nature. Gravitation is one of them. Gravity, the phenomenon whereby objects possess weight, is a universal feature of the world. All things weigh something, from cement mixers to fleas to elementary particles; conversely, no material object weighs nothing. This universality of weight is no accident. It is bound up in the ultimate nature of matter, part of the very meaning of the term. Short of the apparent weightlessness experienced by astronauts, if you don't weigh anything you are not made of stuff.

Gravity is a force of attraction between objects. My weight is the force of attraction with which the Earth tugs at me. Less obvious but also true is the fact that at the same time I am attracting the Earth . . . and the table beside me, and the pencil upon it, and, in general, all things. Everything attracts everything—that's gravity.

Electricity is the other of the two forces that make up the second cosmic number. Like gravitation, the electric force is ubiquitous, and it is fundamental. On the other hand, electricity is not the same thing as gravitation—it is very much stronger, for instance. The second cosmic number describes how much stronger. It compares the electric and the gravitational forces between two things. Between which two things? Between two of the most fundamental things known: two subatomic particles, the electron and the proton. The second cosmic number is *the electrical force divided by the gravitational force between an electron and a proton.*

It's easy to do the arithmetic. But the answer one gets is remarkable: *The first and second cosmic numbers are nearly equal.* That is the mystery. That is the remarkable agreement.

From the day that agreement was first discovered it attracted a good deal of attention. Scientists immediately noticed that while the second of the two numbers is fixed and invariable, the first grows steadily larger as the millennia roll by. After all, it merely measures the age of the universe, albeit in unfamiliar units; ages back in the past it was smaller, ages into the future it is destined to be larger. For this reason the analogy of the train is a good one. The question is why the great red X painted beside the track happened to be even remotely close at the moment someone inside chose to look.

For after all, those two numbers deal with completely different things. What does the ratio of two forces have to do with the age of the universe? What relation could there be between electricity and gravitation on the one hand, and the size of the proton, the speed of light, and the history of matter on the other? These disparities made it difficult to even try to think of an explanation for the agreement. It was like discovering one day that the population of Miami equaled the number of craters on the Moon, and then having to come up with an explanation. Scientists simply did not know where to begin. A further difficulty was that both numbers are so very large. It was hard to see why either should be so gigantic, let alone both. Their very size was intimidating. Certain brave souls ventured forth over the years, proposing various explanations, but none of them seemed to make any sense. The mystery remained unresolved. Nobody knew what to do.

It was not until 1961 that Dicke succeeded in unraveling the matter. He unraveled it by dissolving it.

Dicke dissolved the mystery by showing it was no accident that someone glanced from the train as it was passing the mark. Just the opposite: The mark had reached out and influenced the situation. It had drawn the glance. The X had invented curiosity within the train as it passed by.

He was led to this view by a remarkable discovery he had made while turning over in his mind an apparently unrelated question, that of the lifetime of a star. Stars appear eternal from our limited viewpoint, but they are not so; in truth they are formed, age, and then die. Dicke's question was how long the process took. It could not be answered by looking through a telescope, for stars age too slowly—it would be like trying to see a tree grow with the naked eye. On the other hand Dicke did not need any observations, for he knew how to answer the question using mathematics. He derived a formula.

His formula was based on the physical principles governing stellar structure. On the one hand, a star is held together by gravitation. On the other, it is maintained at great heat by nuclear reactions, liberating energy through Einstein's famous relation $E = MC^2$. Still a third factor is the rate at which that heat flows outward; this, in turn, is determined by processes impeding the flow, primarily electrical in nature but also depending on the properties of the subatomic particles past which it flows.

Skywatchers know that faint stars cannot be seen directly. Look at one squarely and it remains invisible. But peripheral vision has greater acuity: Direct your gaze off to one side of the star and you will find it. Averted vision, amateur astronomers call the technique. Similarly, there are times when it pays not to think too literally. Once in a while the essence of a situation can be discerned only through a mental version of the same trick. Such analogical reasoning is in fact a highly creative skill, and it is regularly employed by good scientists. This was one such occasion. Dicke backed away from all the details of stellar structure, and he studied the subject in the most general possible light. What were its basic ingredients?

Well . . . there was gravitation. And there was electricity. And there was the speed of light and the properties of sub-atomic particles.

But those were the ingredients of the two cosmic numbers.

Dicke played a game. He worked through the theory of the aging of a star and derived a formula for its lifetime. Then he compared that lifetime to a jiffy. He had created yet a third cosmic number, *the lifetime of a star in jiffies*. It was similar in spirit to the first, but referred now to stars rather than to the universe as a whole. Significantly enough, the answer he got was equal to the second cosmic number.

As things then stood, Dicke was facing a peculiar situation. On the one hand, nothing he had done explained the agreement between the first and second cosmic numbers. On the other hand, he had stumbled on a reason for the second to agree with yet a third. It was, of course, an intriguing sequence of equalities. If in some way he could prove that the first and third numbers should agree with one another, he would have succeeded in resolving a literally decades-old enigma. But what could they possibly have to do with each other? One measured the age of the cosmos, the other the lifetime of a star.

Why should the age of the universe equal the lifetime of a star?

Dicke realized the explanation could be found in the Anthropic Principle. He thought in terms of cosmic evolution—of a grand, slow shift over the entire history of the universe. He returned to the crucial role stars had played in making it a fit habitation for man. As recounted in the previous chapter, when the cosmos began it was composed of hydrogen and helium, elements utterly unfit for life. Stars then transmuted them into more suitable material. Hoyle had asked how it was done—how the nuclear reactions that accomplished this had run. Dicke now asked a different question. He asked how long that process of preparation had taken. The answer was that it had required the full lifetime of a star. Not until a complete generation of stars had run through their histories did there exist a sufficient abundance of those chemical elements so crucial to life.

Thus the epoch of man in the cosmos was delimited. We could not have arrived on the scene too early. Life could have arisen only after a certain stretch of time had elapsed, a period of preparation equal in length to the stellar lifetime. "Carbon is required to make physicists" was how Dicke put the matter; and once this carbon was available, physicists could not be too long in coming.

It is, conversely, no accident that physicists did not first measure the two cosmic numbers far in the future, when the age of the universe will be thousands of times greater than the lifetime of a star, thus spoiling their agreement in the opposite sense. So far in the future there will be no stars. All will have gone out—and physicists do not need carbon alone. They also need the life-giving heat and light emitted by the Sun. Without brightly shining suns the Earth and every other planet would quickly freeze, so bringing life to an end. Stars, though, are formed from interstellar clouds, and there are only a limited number of these available. The stellar population is not capable of reproducing itself indefinitely. Once the last cloud is used up, the last generation of stars will have been born, and the end of life in the cosmos will not be long in coming. And so it is that the Anthropic Principle demands that we live in an epoch in which the age of the universe equals the lifetime of a star.

Here is a railroad track. On it is a train. Ages past that train set forth upon a great journey, a voyage of inconceivable magnitude; its track twisting and turning, winding through forests, valleys, and mountain passes, skirting great rivers, prairies, mountain lakes. Vast is that journey and far-distant the destination of the train, but babies are not born upon it, do not grow old and die. Nothing lives within, for it is sterile. Inside, an initial generation of stars brightly shines.

Now the train approaches a great red X painted on the ground. Something new appears; something tiny, wet, and squirming in the mud. A mat of algae floats upon the ocean surface, diatoms sparkle in the depths. For ages a perfect silence has reigned on land, broken only by the soughing of the wind, but now a faint scratching is heard as lizards crawl from stone to stone. As the train rolls on, song breaks forth inside, song and the sound of laughter. Someone is talking loudly. A bell rings.

A hole appears in the train's side. A drill bit pokes through and is quickly withdrawn; an eye peers out. Other holes appear, hundreds of them. Other eyes are clapped to these holes as people jostle one another aside for a look. Arguments break out. Someone notices a mark painted upon the ground; notices, wonders, thinks a bit, and comes upon the explanation.

Now the train has passed that mark. All the stars within it have gone out. Darkness falls, the stygian, unrelieved blackness of the cave. The cold deepens. Snow falls, lakes and oceans freeze into solid blocks of ice. It is 200 degrees below zero. It is 400 below. No sound emerges from within. And still, filled with blackness, filled with corpses, filled with bits of paper covered with strange marks, the Train with No Engineer lumbers on, gently lurching, softly rocking from side to side.

3

Balance

"Mommy, Mommy, why are we wearing shoes?"

"Because everyone wears shoes, honey."

"But why does everyone wear shoes? It's not winter-time—it's warm outside. We don't need them today."

"We wear shoes because that's the polite thing to do. You can take off your shoes at home or in the backyard, but here in the shopping mall it is customary to keep your feet covered."

"Mommy, Mommy, why is there air?"

"Because you need air to breathe. If there weren't any air you wouldn't be here to know the fact. Now for God's sake be quiet for a moment so I can think!"

It takes no great wisdom to spot a difference between those two replies of the beleaguered mother. The first is an explanation. It gives reasons. But the second—an anthropic sort of reply—does not. The Anthropic Principle unquestionably tells "why" the Earth possesses an atmosphere. But it does not tell *why*; it does not explain. The normal explanatory sort of reply to the child's second question would begin with an account of the early history of the Earth, at which time it did not have an atmosphere, then would pass on to the release of gases from the planet's deep interior, the gradual accumulation of those gases about the planet to form a protective layer, and the means whereby the appearance of life altered their composition to their present form. Such an accounting gives understanding. It gives insight. It tells a story. The Anthropic Principle does none of these.

On the other hand it is certainly true: utterly true, tautologically true; and in many instances it provides valuable insight. Dicke's work is an example of what it can do. Like Hoyle, Dicke never explicitly referred to the Anthropic Principle—Brandon Carter had not yet invented the term. To my mind, Carter's coining of that term marked a turning point. It was more than just a matter of naming something. He clearly and unambiguously expressed what till then had been a loosely defined, unarticulated set of ideas. At the same time he carried those ideas further, going beyond Hoyle and Dicke in finding many other such anthropic "explanations"—let us keep those quotation marks—for various facets of the physical world. Others soon followed him in this program, and as a result of their work much understanding has been gained. Carter's efforts set the stage for much that was to follow.

Carter also formulated an important distinction between what he named the *weak* Anthropic Principle and the *strong* Anthropic Principle. Both come into play in responding to the child's question "Why is there air?" The weak principle tells us why we live upon the Earth, which possesses air, rather than upon the Moon, which does not. We live here because "here" is habitable; we don't live there because "there" is uninhabitable, by us and by any creatures remotely like us. The weak principle states that *humanity can exist only in a habitable environment.* But the strong Anthropic Principle goes further. It states that *a habitable environment must exist.* It states that on at least one planet of the cosmos, air must exist; that at some obscure location in an ordinary galaxy, at some otherwise unremarkable point in the grand scheme of cosmic evolution, a planet must be found wrapped in gasses of precisely the composition required by humans. Since we exist, that gas exists.

Carter's distinction comes into play in comparing the two coincidences discussed in the two preceding chapters: that of the cosmic numbers on the one hand, and that of the three nuclei on the other. The weak Anthropic Principle relates to the first coincidence, the strong principle to the second. Are these coincidences of the same sort? On the surface it may seem that they are, but in truth they are not. The first, in fact, is not really a coincidence at all. After all, if you wait long enough, a train is guaranteed to pass some X marked beside the tracks. Dicke's point was simply to show that the mark indicates the habitable portion of the track. We live upon the Earth rather than the Moon; similarly and as required by the weak Anthropic Principle, we live alongside the X because everywhere else is unsuitable. Given this understanding, all elements of coincidence vanish from the first example.

But the second example is different. The resonances between helium, beryllium, and carbon relate to the very existence of habitable environments in the universe, and as such they fall under the aegis of the strong Anthropic Principle. Had those resonances not occurred, no location would have been habitable—not the Earth and not the Moon, not anywhere within our Milky Way Galaxy nor within any other galaxy, nor indeed at any other epoch of cosmic history.

Furthermore, those resonances really are coincidences. They are genuinely remarkable strokes of luck. As emphasized above, the Anthropic Principle provides no explanation for anything, and no amount of anthropic reasoning can explain these coincidences away. The point can be seen most vividly with the help of Leslie's analogy, mentioned in the Prologue, of the firing squad. A court has sentenced a man to be shot at sunrise. Early this morning he was dragged before the firing squad. He stood before it blindfolded, the commander gave the order to shoot . . . and every last one of the rifles misfired.

Had even one functioned properly he would not be alive today. The bare fact of his existence therefore proves they must all have misfired. That's the strong Anthropic Principle in operation. But clearly it is not enough. It is not enough to explain why so very many coincidences all occurred at once.

Coincidences covered by the weak Anthropic Principle are of no interest to me in this book. While interesting in their own right, they pose no mystery. But coincidences covered by the strong principle are another matter. They do pose a mystery: a great and profound mystery, I believe, and one of immense significance. The analogy of the firing squad is a good one in this regard. Here is another rifle in that squad that failed to fire.

Somehow, an insect seemed to have gotten into my canteen.

The thing buzzed at me as I lifted it to my lips. Six of us sat resting there on the high mountain ridge that afternoon—six people and one dog, a German shepherd as I recall. We were on a mountaineering vacation in the Rockies. We had set off early that morning under clear skies, but the weather had soon clouded over and now distant thunder echoed ominously among the peaks. Best to lie low for a while and see how things developed.

Gradually the sweat dried on my brow. Lowering the canteen I shifted my back against the unyielding granite and let my gaze wander across the magnificent vista.

In the foreground the precipitous slabs of the ridge led downward into a jumble of massive boulders. At the mountain's base a clear blue lake lay amid the scree, an icefield by its side. Faintly I could make out the sound of a frigid stream tumbling down to yet another lake, set among soft meadows. In the far distance the terrain lifted again into peak after peak, icefield after glacier. The sky was a uniform heavy gray.

I lifted the canteen to my lips again. It buzzed more viciously. And at that very moment a woman seated beside me rose to her feet. As she did so, every single strand of her long hair lifted upward and stood straight out from her head.

I stared at her in stupefaction. In an instant, with a yell, all of us leaped to our feet and vaulted over the edge of the ridge. We tore down the mountain. Stretches we had negotiated anxiously on the way up we now positively jumped over. I reached the top of a long slab and actually ran down its steep slope. All the while the dog, barking and yelping excitedly, ran between our feet and nipped at our heels—the damn thing thought it was a game we were playing. I lifted my hand above my head: My fingertips tingled. Nervously I jerked it downward. When I stood upright, my ears buzzed and felt as if a cloud of moths were fluttering against them; I ran bent double.

A terrifying flare burst about me, a concussion, a single monstrous *crack* so overwhelming it was beyond sound. I jumped involuntarily and clapped my hands to my head. Leaning against a boulder I looked at my companions. We exchanged glances with scared eyes. Nobody had been hurt.

Lightning is a flow of electric current, no different in kind from the stuff that flows in the wire when the lights are switched on. In both cases the flow is pushed along by the force of electricity. As discussed in Chapter 2, electricity is ubiquitous, fundamental, and strong—very strong.

Even the terrible power of lightning represents only the barest hint of the force of electricity's true potential. That potential, however, is never realized . . . and this is the next of those marvelous coincidences upon which our existence depends. Indeed, the electrical force spends its time in hiding, a dragon crouching in a cave. When the lights are turned on, the barest tip of the monster's nostril has poked forth; when lightning bursts, a mere few inches of its head. Not once in history has the beast fully emerged to wreak its havoc.

How does the beast remain so perfectly in hiding? It is tempting to suppose it does this by being rare, an aberration in the normal course of events. That would jibe with what appears to be a relative paucity of electrical phenomena in nature. For after all, most of its manifestations are artificial: electric heat, television, and the like. While prevalent in industrial society, its workings appear a peripheral aspect of the natural world, and aside from lightning its only instance would seem to be the mild shock one experiences from time to time when walking across a deep carpet and reaching out to touch something.

But this is an illusion. In truth, electricity is everywhere. As a matter of fact, my experience on the mountain was literally crammed with evidence of its ubiquity. Electricity is present in hair—that's why the woman's responded so dramatically as she rose to her feet. Electricity hides in canteens—that's why mine buzzed. It floods fingertips and ears and sets them tingling, inhabits stones and boulders and so attracts lightning to them, fills storm clouds and the very air itself and so creates the burst. Had I been free of it I would have been entirely safe that day upon the mountain ridge. But I, like every other object in the universe, am composed of the seeds of lightning.

This being the case, it is all the more remarkable how perfectly the beast remains in hiding. It does this not by being absent but by striking a balance, a delicate and precise cancellation of opposing tendencies. Electricity walks a tightrope, and the perfection of its balance is quite extraordinary—better than one part in a billion billion. Our lives depend on it. If this balance were to be upset, even for an instant, each one of us would burst apart in an explosion. As would each tree, each blade of grass, each planet, and each star.

Each tree, each blade of grass, each planet, and each star is made of atoms. Atoms in turn are made of electrons, protons, and neutrons. These three subatomic particles are the building blocks of all things, and there is not a single thing that does not contain them all. Put them together in such and such a way and you have built a piece of Shakespeare; put them together in different proportions and in a different arrangement and you have built a piece of wood. The neutron, being electrically neutral, plays no part in electrical phenomena. But the remaining two most decidedly do. They are the seat of the electrical force.

Take a piece of silk and two glass marbles. Rub the marbles with the silk and place them on a table. They will tend to roll apart. Now take two rubber balls and rub them with a bit of fur. They too will roll apart. On the other hand the glass balls will attract the rubber balls. They will roll together.

Rubbing the glass balls has placed a positive electric charge on each. The fact they rolled apart shows that two positive charges repel one another. Rubbing the rubber balls, on the other hand, induced a negative charge. Remarkably enough, these also rolled apart. The moral is easy enough to draw: Like charges repel one another. Similarly, the coming together of the glass and rubber balls demonstrates that opposite charges attract. The same is true at the microscopic level. Electrons are the carriers of negative charge and protons of positive charge; electrons repel electrons and protons repel protons, but electrons and protons attract one another.

Neither of the two balls attracts me, however, nor any other electrically neutral object. Because my body is literally flooded with electric charges—indeed, it's built of them—it has only one way of maintaining this immunity: by containing equal numbers of each type of charge. If, say, one hundred protons are to be found in my fingertip, one hundred electrons will also be found there. That balance is inherent in the nature of matter: The plus just cancels the minus, adding up nicely to zero.

That is the balance . . . or rather, half of it.

It is half because in order for matter to remain electrically neutral in this way, the charges of the electrons must exactly balance those of the protons. These charges have been measured experimentally and it was found that they do. But what would the universe have been like if they did not?

If the electron carried more charge than the proton, the cancellation would no longer be perfect. Objects would still contain equal numbers of each particle but the charges would no longer add up to zero. Rather, they would add up to a net minus. Every object in the universe would be charged negatively—and like charges repel one another. The table in the corner would repel the lamp upon it; the tree outside the window, the car parked beneath. My left and right hands would possess a net negative charge; as they were shoved apart my arms would extend outward. My feet would repel my head, my blood my bowels, and, in general, my entire body would be forced to expand. As would the lamp, the table, the tree, and the car. A universal repulsion would result.

Its strength would depend on the degree to which the charges of the electron and proton differed: small differences, small repulsion. On the other hand, so powerful are electrical forces in general, and so enormous are the numbers of subatomic particles within even the smallest of objects, that even very small differences would lead to very great forces. A mere 1 percent offset between the charge of the electron and that of the proton would lead to a catastrophic repulsion. My arms would not just extend outward; they would rip from my body, blast away from the Earth, and actually fly off into space. As they rose upward (followed by teeth, hair) they themselves would distend and burst. My entire body would dissolve in a massive explosion. So too the ocean, which would rise from its bed in a rush, flashing into steam—cold steam. The mountains would rip from the fabric of the continents to hurtle upward, dissolving into pebbles, the pebbles into dust. The very Earth itself, the planet as a whole, would crack open and fly apart in an annihilating explosion. The same would happen to every other planet, and to the Moon— all moons—and to the Sun and stars and, indeed, to every material body in the universe.

That is what would happen were the electron's charge to exceed the proton's by 1 percent. The opposite case, in which the proton's charge exceeded the electron's, would lead to the identical situation—for all things would then be charged positively, and like charges repel one another no matter what their sign.

No . . . 1 percent is too great a discrepancy. How precise must the balance be? How accurately must the charge of the electron be matched with that of the proton before such disasters can be avoided? The question is easy to answer—a good high-school student could do it—but the result one obtains is impressive: Relatively small things like stones, people, and the like would fly apart if the two charges differed by as little as one part in 100 billion. Larger structures like the Earth and Sun require for their existence a yet more perfect balance of one part in a billion billion.

Evidently nature has adjusted things to better than that accuracy. Otherwise, physicists would not be here to learn the fact—neither physicists nor anyone else. Once again, lacking such an extraordinarily precise adjustment of its microscopic properties, the cosmos would have been uninhabitable, uninhabitable throughout all of space and time. It would have been unknowable, invisible. Once again the strong Anthropic Principle "explains" all . . . and once again, the explanation it offers ultimately fails to provide an accounting.

4

Blue-White Planet

I have already mentioned Lawrence J. Henderson, professor of biological chemistry at Harvard, and his remarkable 1913 work *The Fitness of the Environment.* Henderson's was a seminally important book. It set forth the problem with which I am here concerned, a problem that has grown only more acute with the passage of time.

Also worth noting about Henderson's book is its title: It is a play on the well-known concept of the fitness of organisms to the environment. At the very outset Henderson wanted to emphasize the essential link between his concerns and Darwin's theory of evolution. Darwin's theory explained how organisms came to be fitted to their environment—but Henderson's essential recognition was that not all environments could be fitted to. Fitness was a coin with two sides, and evolution dealt with only one of them. "Darwinian fitness is compounded of a mutual relationship between the organism and the environment," he wrote, "and *the actual environment is the fittest possible abode of life*" [emphasis added].

The year 1913 is long gone now, and our knowledge has advanced enormously since then. Most of modern physics and astronomy lay far in the future in Henderson's day. The bulk of the arguments with which we are now concerned were not available to him and do not appear in his book. On the other hand, many of his arguments still ring true. In particular, his treatment of the properties of water was right on target.

Ours is the blue-white planet. Those are its predominating colors as revealed by photographs from space: the exquisite blue of the oceans, the breathtaking white of the clouds. In contrast, the continents are positively difficult to spot in these photos, their outlines so often obscured by clouds and their colors so dun. To my mind the Earth is the most lovely of all the planets of the solar system—and it is worth noting that its beauty as well as its predominating colors both come from the same cause: water, water in the oceans and water vapor in the clouds. We named our planet wrong.

Water is equally predominant in living things. Human beings are roughly 60 percent water. Furthermore, much of the remainder is provided by bones; take away these supporting struts and the percentage grows yet higher. Punch a hole in a person and blood oozes out—and blood is essentially water: In terms of composition it is remarkably close to seawater, and its most visible characteristic, its brilliant red color, is provided by the addition of quite minor amounts of hemoglobin to the mix. Nor is blood the only aqueous component of the body. Each individual cell contains great amounts of the stuff. Of course people look dry, and feel dry to the touch, but this is an illusion, for what you see when you look at a person is skin, and the outer fraction of an inch of skin is composed of dead cells. As for other organisms, the situation is identical. Cats, trees, spiders, it makes no difference: The percentage of body weight taken up by water varies, but this percentage is never small. There is not a single organism known that makes do without that liquid. To be alive is to be wet.

Every organism is a factory, maintaining its existence by virtue of an immensely complex interworking of various parts. As in all factories, goods must be transported from one place to another. Oxygen inhaled into lungs must be carried to the cells; carbon dioxide, a waste product of cellular metabolism, must be carried back to the lungs to be exhaled. Nutrients ingested as food into the stomach must similarly be transported about the body, as must waste destined to be deposited into urine. From the soil, the root system of a tree gathers nutrients, which then are carried up its long trunk and to the farthest tip of each twig; glucose, manufactured in leaves by photosynthesis, is likewise spread throughout the plant. In these and countless other ways every living being is pervaded by a network of transportation pathways. These pathways—these conveyor belts—are made of water.

Why water? Why not some other liquid? Substances are not transported about the body in solid form like so many pebbles washed downstream in a creek. They are *dissolved into* the liquid that transports them, and they are carried from place to place as solutes. It turns out that water is the best substance available for this job. In Henderson's words, "As a solvent there is literally nothing to compare with [it]." No other commonly occurring liquid possesses anything remotely approaching the ability of water to dissolve things.

When a solid is dissolved in water, each component molecule is removed from the solid configuration and released into the liquid environment. Water molecules cluster about these solute molecules and surround them. This clustering is essential. It is the mechanism whereby they are isolated from their fellows. If the solvent molecules did not do so good a job of isolating those of the solute, they would be free to seek one another out and recombine into large-scale aggregates—to solidify again. But why do the water molecules so cluster?

It is often the case that each molecule of the dissolved substance is dissolved as an ion—a form in which one or more electrons have been removed. But these electrons have negative charges. Thus the dissolved ion has a positive charge. As for the H_2O molecule, it too has electrical properties, though they are slightly more complex. Its structure is diagramed in Figure 7. In this configuration the oxygen atom

Water molecule

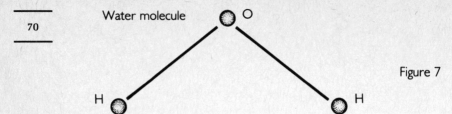

Figure 7

tends to attract the electrons belonging to the two hydrogen atoms. They migrate toward it, and as a result the oxygen becomes charged negatively. Similarly the hydrogens, lacking these negative charges, acquire positive charges. The result is not a net charge but a *charge imbalance* within the H_2O molecule, schematically diagramed in Figure 8.

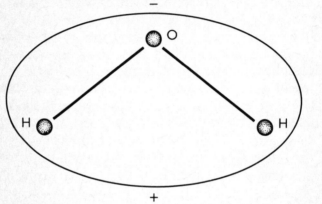

Figure 8

The magnitude of this charge imbalance is known as the molecule's dipole moment.

Water molecules cluster about those of the solute for electrical reasons. Unlike charges attract: The negative ends of the H_2O molecules are attracted to the positive charge of the ion. The configuration is sketched in Figure 9.

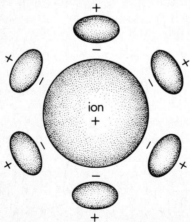

Figure 9

In other situations the dissolved molecule is not ionized but nevertheless possesses a charge imbalance similar to that of H_2O; here too, the clustering arises for electrical reasons.

The more firmly H_2O clusters, the more effective it is in isolating each molecule of the solute, and the more effective water is as a solvent. Water clusters very effectively indeed, and it does so for a simple reason: *It has the largest dipole moment of any commonly occurring molecule.* It is difficult to think of a molecule whose dipole moment even remotely approaches that of this liquid. And as a consequence, water is uniquely suited to its role as a transport mechanism. The ancient alchemists labored without end in search of the universal solvent. Little did they know they were composed of it.

Solar energy converters festoon the surface of the Earth, untold numbers of them covering literally millions of square miles of territory, so ubiquitous we hardly notice their existence. A long-term technological projection? Far from it: That is the present reality. These solar cells are leaves, grasses, and the like; and they are in the business of trapping the energy of sunlight and converting it into a biologically useful form.

Plants use the energy so trapped directly. Animals use it indirectly—by eating the plants. The cow munching in the field is utilizing solar energy via a two-step process. The carnivorous wolf utilizes it through yet a third step, that of eating the cow. We, in turn, are omnivores, and run on solar energy via either a two- or a three-step chain. But omnivores, herbivores, and carnivores alike depend upon vegetable matter for their ultimate source of food. And vegetable matter depends upon its solar cells.

Photosynthesis is the chemical reaction whereby these cells trap the energy of sunlight. In this reaction six molecules of water are combined with six of carbon dioxide to form one of a sugar known as glucose, with the concomitant release of oxygen into the atmosphere. In chemical notation:

$$6H_2O + 6CO_2 + \text{sunlight} \rightarrow C_6H_{12}O_6 + 6O_2$$

glucose

The glucose so produced is a high-energy molecule, and it forms the ultimate basis of all foods. The energy it contains is the fuel on which life on Earth depends.

Note the presence of the H_2O molecule in this reaction. It is essential to it. The photosynthetic process could not occur without water.

There's more. Photosynthesis does not make glucose alone. It also makes oxygen. Photosynthesis, in fact, is the source of *all* the oxygen in the atmosphere, and oxygen is essential to the existence of every breathing animal. Without photosynthesis we would suffocate.

We would also die of ultraviolet radiation. The Sun is a powerful source of ultraviolet, which floods down upon the Earth in great quantities. But just before reaching the surface this deadly radiation is absorbed—absorbed by the protective ozone layer of the atmosphere. Ozone is oxygen: O_3.

Where does the oxygen evolved in photosynthesis come from? It is present in both of the molecules entering into this reaction, carbon dioxide and water. The answer can be obtained by tagging one or the other with radioactive tracers, and finding which causes the evolved oxygen itself to be radioactive. It turns out that water is the source of all the oxygen in the atmosphere.

It's a hot day. I'm drenched in sweat. The dog lies exhausted in the corner, its tongue lolling out. Both of us are doing the same thing: We're cooling ourselves. The dog does this with its tongue, I with the entire surface of my body. Each is covered with water.

It is essential for warm-blooded animals to cool themselves in hot weather. Mammals need to keep their body temperature from getting too high. A fever of a few degrees is cause for alarm, one of 10 degrees fatal. That's a small excursion in temperature compared to the 98.6 degrees that is normal; mammals require an exceedingly effective thermostatic mechanism to survive. And this mechanism relies on water.

How does it work? Both the dog and I cool ourselves by transferring water from the interior of the body to some portion exposed to the air. This exposed water then evaporates.

But energy is spent in evaporating a liquid. The energy so spent is sapped from the body, and because it is sapped the body temperature is lowered.

The so-called heat of vaporization is a measure of the efficiency of this process. It measures the amount of energy spent per unit quantity of liquid evaporated away. The higher this heat of vaporization, the more energy is sapped from the body and the more efficient is the cooling. Water turns out to be far and away the best substance in this regard. There is hardly a substance known to science whose heat of vaporization exceeds that of H_2O. There is not a single commonly occurring liquid whose heat of vaporization comes even close. If we sweated not water but something else, body temperature would be subject to unacceptably great increases in hot weather. Summer would be lethal. Heavy exercise would be deadly.

There is a sense in which the Earth as a whole sweats. After all, the bulk of its surface is covered with water—the oceans. In equatorial regions these oceans evaporate under the blaze of sunlight, and waft into the air as vapor. Thus the tropics are cooled. But the vapor so produced does not remain in the tropics. It drifts away from the equator, northward and southward into cooler regions. There it condenses into rain or snow, and just as energy is spent upon evaporating a liquid, so it is released when the vapor condenses. Thus the colder regions of the Earth are warmed.

Water achieves this moderating effect on climate in yet another way. Every cook knows that energy is required to heat water—that's why you have to turn on the stove. Another way of saying the same thing is to say that the liquid stores energy in the form of increased temperature. And it develops that an unusually great amount of energy can be stored by water in this way. The same quantity of energy applied to an equal mass of rock would raise its temperature by five times as much; an astonishing thirty times as much if applied to an equal mass of lead. The consequence is that the presence of water diminishes the variations of temperature from day to night. The full heat of the noonday sun is absorbed without raising the temperature too much. Similarly, the heat of summer is mitigated as compared with the cold of winter. In contrast, regions devoid of water—deserts—suffer far greater variations, with the 120-degree blaze of the mid-summer day alternating with relatively cool nights, and sometimes even snow in winter.

Were it not for water's immense heat of vaporization, and its ability to absorb great amounts of energy while undergoing only mild increases in temperature, our climate would be bitterly severe. The sunny Caribbean would be an inhospitable wasteland of blazing heat, the mid-latitudes arctic, the poles even colder than they are in reality. Daytime would be scorching, the nights frigid.

Henderson also pointed out that water expands upon freezing. The expansion is not very great—a fraction of a per-cent—and it is tempting to think the practical effects should likewise be small: a little antifreeze in the radiator of one's car to keep the engine block from cracking and that's that. In reality, though, the consequences are enormous, for it means that solid H_2O is buoyant—*ice floats*. Water, in fact, is rare among liquids in possessing this property. Most liquids contract upon solidifying. If water behaved "normally," ice would sink.

But if ice did sink, the environment would be changed radically. With the coming of winter the surface of each lake and pond would freeze over—and the ice, rather than remaining on the surface, would sink to the bottom. Then the next layer, unprotected now from winter's cold, itself would freeze, contract, lose buoyancy, and settle downward. Ultimately the entire lake would freeze into a single solid chunk of ice, so killing all life within it. With the coming of summer the surface would melt, but the bulk of the ice would remain solid, lying on the bottom and insulated from summer's warmth by a thin upper layer of water. This would be true not just of lakes and ponds, but of oceans as well. The blue-white planet would be the icy planet.

In such an eventuality, only those regions whose temperatures never dropped below freezing would contain significant quantities of liquid water. The portion of the globe amenable to life would be confined to a narrow strip lying close to the equator. On either side of that strip, the land- and seascape would be indescribably barren and inhospitable, the lakes, ponds, and oceans remaining largely frozen even in the balmy days of summer. Although it is dangerous to underestimate the adaptability of life, the astonishing ability of organisms to adjust to harsh environments, it is hard to see how the incomparable richness and diversity characteristic of life as we know it could flourish in such a world.

5

Seeds of Life

Throughout Henderson's discussion of water and its properties there is one striking omission, an aspect of the situation he failed to mention. He never wrote about how much there is of it upon the Earth.

It is easy to become blinded to the truly remarkable prevalence of water upon our planet. We are, after all, land creatures, and most of us live out of sight of oceans. A sea voyage would be a good corrective to this general misapprehension. Another is a photograph of the Earth from space. And visible to us all, floating high above our heads, is yet a third—the Moon. If the Moon contained anything remotely approaching the quantity of water found on the Earth, the view we have of it would resemble in miniature that portrayed in photos of our planet: blue of sea, white of cloud. In reality, however, there is not the smallest of ponds, the most meager of brooks to be found on our satellite. The lunar landscape is an utterly barren wasteland of desiccated stones.

The same is true of Mercury, innermost planet of the solar system, which in many ways resembles the Moon. Venus, second planet outward from the Sun, is an inferno, with temperatures exceeding 800 degrees Fahrenheit, no liquid water on its surface, and only minute traces of vapor in the atmosphere. Mars, orbiting just beyond us, likewise contains no oceans, lakes, or streams, and while significant quantities of H_2O may be found there, the Martian environment is so cold that water would exist as permafrost below the surface. Yet farther away from the Sun the environment grows even colder, and the conclusion remains the same. Ice, however, is of little use in the biological makeup of living beings; water must exist as a liquid to be of utility. Furthermore, in order for water to exert its important moderating effect on body heat and climate, the temperature must be such that it regularly alternates between the liquid and the vapor states. Of all the planets of the solar system, only one possesses the twin characteristics of carrying great quantities of water and of lying the appropriate distance from the Sun to have the right temperature.

This being the case, it might seem all the more remarkable that Henderson made no mention of the fact. Is it not yet another example of the extraordinary fitness of the environment for life?

No, it is not. It is utterly irrelevant. An analogy to our situation can be found in the example of an alpine flower high upon a mountain ridge. The ridge is a wasteland of cliffs and barren stones. But set amid these cliffs, nestled among the stones, is a niche—a tiny cleft. That cleft contains a trace of soil and offers a modicum of protection from the elements . . . and there bravely grows the flower.

One might be tempted to wonder why the flower grows in the niche. After all, suppose it had attempted to sprout somewhere out on the barren rock face? But the question is misguided. In reality there is no mystery: The flower is there because "there" is the only place it can flourish. And in a similar way, we live upon the blue-white planet because it is the only suitable environment around. Carter's weak Anthropic Principle explains all: The Earth, our planet as a whole, is itself a niche, a haven set amid cosmic wastes. That is why we are here rather than somewhere else. And that is why Henderson paid no attention to the vast amounts of water upon the Earth.

The Earth rolls onward, carrying us, flowers, and oceans with it. As it rolls it orbits the Sun, which wanders through a swarm of stars. There is no danger of collision with those stars, for we are protected by their great distance from us. Other regions of the universe, though, are not so hospitable. Places are known where the stars lie close together. Figure 10 shows one: a great globular cluster of stars, hundreds of thousands of them, and all relatively tightly packed. Such clusters are common—and the densest are utterly inhospitable to life. The stars within them suffer close encounters with one another too regularly for their planets, if any should exist, to remain in circular orbits about their suns. Another such in-

Figure 10: The great globular cluster in the constellation of Hercules—a possible example of a bad niche. *Palomar Observatory*

Figure 11: Another possible bad niche: the nucleus of the Andromeda Galaxy. *Mount Wilson and Las Campanas Observatories, Carnegie Institution of Washington*

hospitable location is shown in Figure 11: the center of a galaxy. These centers, or nuclei as they are known, are regions of complex, violent activity and tightly packed stars. I would not expect to find life anywhere within them. In contrast, the Earth lies far out in the outskirts of our Milky Way Galaxy, in sparser, safer regions. In a grand, a cosmic niche.

The concept of niche can be broadened yet further, to include the possibility of its existing not just in space, but in time as well. That is what Dicke did. Only after a significant quantity of carbon had accumulated in the universe, only after this great, slow transmutation of the stuff of the world had progressed, was the stage set for the emergence of life. But all this took time. And so it is that the cosmos we observe about us is immensely aged. "Now" is our niche.

*   *   *

High upon a mountain ridge, tiny white fluffs are blown before the wind. One passes close to me; grabbing it in my hand I find it is a tangle of fine white threads. From its base dangles a tiny seed.

Most of these seeds are scattered by the wind across the barren rock face. There they lodge, ultimately to wither away and die. One, however, by chance is blown into a tiny cleft between the boulders, a cleft in which some soil has accumulated. There that seed comes to rest. Several months later, were I to return this way, I'd find a flower there.

That is how the flower finds its niche. That is why it grows there. There's a question, though. If the Earth as a whole can be thought of as a niche, a niche in the solar system, a niche in the Milky Way Galaxy—how did life find this out? If "now" is a niche, how did biological organization arise at the right time? What are the seeds, not just of flowers, but of life itself? How did they get here?

In the early years of this century the Swedish chemist Svante Arrhenius actually did propose that seeds floated among the stars in the form of tiny spores. The hypothesis was dubbed panspermia. Arrhenius knew that spores were remarkably hardy, and he speculated they could withstand the rigors of interstellar travel. Since then we have come to realize that space is far less benign an environment than had been thought, but the notion remains conceivable. More recently the hypothesis of artificial panspermia has been advanced: that ages past some highly advanced civilization sent forth in all directions a flotilla of automated spaceships, each carrying primitive microorganisms with which to infect any planet it encountered. We might be descended from extraterrestrials.

It's not impossible, and if true would answer the question. Life would flourish upon the Earth for exactly the same reasons that flowers sprout from clefts. But most scientists do not accept the panspermia hypothesis. By far the most commonly accepted theory is that life on Earth arose, not from some kind of seed left behind by a prior generation of living things, but from nonliving matter. Life is thought to have originated right here.

SEEDS OF LIFE

Research is actively under way on how this might have come about. Laboratories are dedicated to the subject, conferences held upon it. A bolt of lightning, or a cosmic ray, or ultraviolet light from the Sun is thought to have struck a complex molecule in a tide pool, or on the surface of a tiny airborne particle, or in the ooze of mud in a bog . . . and triggered the formation of something we would be willing to call alive. There's a big gray area in there, and the actual mechanism is much in doubt. But scientists agree there was nothing supernatural involved: The origin of life upon the Earth came about as the result of purely natural causes. Although we do not understand how nature did it, there is not the slightest doubt that it was indeed nature that did it. The first organism was constructed of naturally occurring molecules with the help of a sudden burst of energy from some outside source. Molecules and energy were all that was required for the construction of the first living organism.

Matter and energy—these are the seeds of life. And matter and energy are everywhere. That is how the seeds got here. They had been here all along.

And as I listen to the faraway murmur of surf upon the beach, as balmy breezes fan me with their coolness, as in the dead of winter I am warmed by sunrise—as I contemplate this immense good Earth, it strikes me that the entire planet might be considered a seed, a single gigantic seed that has borne fruit. And visible to me also is a second seed, but one that failed to germinate: a faint white crescent floating in the blue dome of sky arched high above my head—the Moon. So far as the generation of life is concerned, the Moon's only problem is size. It is too small. Had our satellite been larger, its stronger gravitational pull would have held to it an atmosphere, that atmosphere would have kept its water from evaporating away into space . . . and alien flowers would have sprouted from the lifeless stones.

6

Rules of the Game

"East or west, home is best." The Earth is our home, and like all fond homebodies comfortably ensconced before the fireplace, we have a tendency to overrate its virtues. But the weak Anthropic Principle shows that the great amount of water on our planet is utterly irrelevant. We should not be misled by the appropriateness of the Earth's temperature. To dwell upon such matters would be akin to a flower's expostulating upon the virtues of its cleft.

The real enigma is something else. It is not that niches make good places to live. It is that niches exist at all.

My concern in this book is with a mystery, a great and profound mystery I believe, and one of immense significance: the utterly unexpected habitability of the cosmos. But it is important to get the mystery straight. It has nothing to do with the Earth. Had our planet not been hospitable, life could have developed somewhere else. Had the Earth been smaller it would have resembled the Moon; had the Moon been larger life might have originated there. Had the Sun been more luminous, the Earth would have been intolerably hot; Mars, though, would have been warmer too, and if, as we believe, it does contain significant amounts of H_2O, that planet would have provided a fine environment for habitation. Dwellers on the Red Planet would have surveyed their skies with complacency, grateful not to have been consigned to the scorching wastes of Earth.

It makes no difference where your niche is. What matters is that niches exist—and this is the mystery. The habitability of an environment is compounded of two parts. The first, included under the aegis of the weak Anthropic Principle, has to do with being in the right place, and although certainly important there is nothing particularly remarkable about this. It is within the second of these two parts, covered by the stronger form of the Anthropic Principle, that the problem arises: with the existence of niches. If water did not possess the remarkable properties outlined in Chapter 4, there would be no habitable environments. Were it not for the abnormally large dipole moment of that molecule, its participation in photosynthesis, its unusual expansion upon freezing, immense heat of vaporization, and remarkably great ability to store heat—were it not for these things, humanity could not have developed anywhere: not here on Earth, not on the Moon, not anywhere in the universe. Nor could grasses, trees, mosquitoes.

In the hearts of the red giant stars a gigantic and unlikely coincidence is at work. Were it not for that coincidence, neither we nor any other life form could have come into being. Neither the Earth nor Mars would have been a fit location for life—they would not even have existed. Stars such as the Sun would have existed, pouring their luminous energy outward into space, and the giant planets such as Jupiter and Saturn, which are essentially gaseous spheres, would have existed too. But they would have been composed solely of hydrogen and helium, and could not have supported any form of biological organization.

Had the charge of the proton not precisely balanced that

of the electron, the cosmos would have been yet more inhospitable. No coherent structures of any sort could have existed—no Sun with its life-giving heat and light, no rocky planets such as Earth, no gas giants like Jupiter, no stars or planets elsewhere in the universe, and indeed no separate objects of any sort, nothing distinguishable from its environment that might be termed an organism. The cosmos would have consisted solely of a uniform and tenuous mixture not so very different from air.

The matter of collisions between stars is a bit more complex. Plenty of regions exist in which the packing of stars is dangerously dense; plenty also exist that are safe. To this extent the matter is one of life flowering in the right place. On the other hand, there is a serious question of why such niches exist at all; for were the overall distance scale smaller, were stars significantly closer together everywhere throughout the universe, no location would be safe. Stars are arranged into galaxies, and galaxies have a certain characteristic size—but if these galaxies were less roomy, if they packed into the same volume of space a greater number of stars, the cosmos would be uninhabitable. Furthermore, the cosmos itself has certain characteristic dimensions; were it smaller galaxies would overlap, and the resulting stellar crowding would lead to the same result. As outlined in Chapter 10 there is a very serious difficulty in understanding how the universe came to reach anything even remotely approaching its present great size.

None of these difficulties can be avoided by seeking out the appropriate location. Either reality is fit for life or it is not. There are many planets, but only one universe. You don't get to shop around.

In Henderson's time the anomalous and life-giving properties of water were entirely inexplicable. No one knew why that liquid behaved so differently from others. In our time an understanding of this question has been achieved, but for decades the extraordinarily precise balance between the electron and proton charge proved equally difficult to comprehend.

That balance is quite mysterious. The electron and the proton differ from one another in all other respects. The proton, for instance, is far larger than the electron. Also it is heavier; the comparison is that between a person and an acorn, pretty nearly. The two particles' magnetic properties differ wildly. And finally, the proton participates in a variety of elementary particle processes involving the so-called strong interaction, for instance the fission and fusion reactions that power nuclear weapons, but these are reactions the electron sedulously avoids. Only in this one regard, charge, do these two particles resemble each other—and surprisingly enough, the resemblance is perfect.

For decades it remained a peculiar situation. The equality of the electron and proton charges appeared to play essentially no role whatever in the microscopic realm, and it was very difficult to understand. There was a general feeling that were those charges to differ slightly, the structure of these particles could hardly be affected. Nor would that of atoms and molecules. Only large objects would be affected. On the other hand, it was presumably in the microscopic realm that those properties arose. Whatever regulated so precisely the electron and proton charges presumably lay buried within them, and had nothing to do with the existence of stones, humanity, and stars. But for generations the reasons eluded us.

And yet what immense strides were made in other areas during that period of time! Most of modern science was developed within it: relativity and cosmology, the discovery of DNA and the mechanism of heredity, elementary particles, the computer, artificial intelligence, voyages to the Moon and planets . . . It is hard to think of a single area of knowledge that was not transformed beyond recognition while that one mystery remained unsolved.

Faced with such a situation, faced with decades, generations wasted in fruitless attempts to storm that impregnable fortress, certain people felt compelled to take a radical step. They advanced the possibility that here we were dealing with a problem with no solution. *Maybe there was no explanation.* No explanation, that is, lying in the natural realm.

Could it be that suddenly, without anybody's looking for

it, evidence had been found of some supernatural Agency at work in the world? The question was doubly thrust forward by the immense significance to our being of that charge balance. Was it possible that this Agency, out of concern for our existence, had stepped into the normally orderly workings of things and altered things to our benefit? Religious people would call it God. Others would not be willing to take so great a step . . . but they would keep the capital *A* on Agency all the same.

It is a matter of taste how one deals with that notion. Those who wish are free to accept it, and I have no way to prove them wrong. But I know where I stand—I and every other scientist. I reject it utterly. I will have nothing to do with it. My conviction is that the world obeys laws, the laws of nature, and that nothing can ever occur that stands outside those laws. To suppose that as they travel from one point to another the electron and proton obey the principles of quantum mechanics, that as they attract and repel each other they follow the dictates of electricity, that their reactions are governed by laws of elementary particle physics—to suppose these things, which are demonstrably true, but then to go on and suppose that in one particular regard these particles stand utterly outside the laws of nature . . . I find this nonsense. The notion that all liquids but water behave according to physics is a notion I cannot accept.

I am sure that ultimately we will come to an understanding. I am sure that some day an explanation will be found, and this explanation will be couched in terms of the laws of physics. Call it faith, if you will. It is my faith, and it is the faith of every scientist.

In recent decades the remarkable properties of water have been entirely comprehended in terms of the principles of quantum mechanics. In recent years the very exciting possibility has arisen that the equality of proton and electron charges may be explained by the so-called grand unified theories of elementary particles, described in Chapter 11. These theories are still highly speculative, and their validity is largely untested. Perhaps they will stand the test of time, perhaps not. No matter: If they fall something else will come to take their place. There may be agencies, but there are no Agencies.

A man and a woman sit hunched forward over a board. The board is ruled off into squares, and upon these squares are scattered a variety of small plastic pieces.

The woman reaches out, picks up one of those pieces, and moves it to a new location on the board. She leans back. Now the man's turn: After some thought he moves a different piece. I am intrigued. What's going on? Studying the game I soon come to some general conclusions. To begin with, I notice the players alternate, first one moving a piece and then the other. Also, the pieces are of two classes—two colors— and each player moves only one of the two classes. Third, each type of piece seems to have definite rules governing its motion: The smallest move only short distances, the larger ones with pointed tops any distance but always diagonally, and those shaped like castles forward and back.

In this way, without ever having been taught them, I learn the rules of the game of chess. But now another game:

High above our heads the grave Moon swings about the Earth, gradually shifting from crescent to full to new phase. Elsewhere a mighty star explodes as a cataclysmic supernova. A quasar flares and churns. Lightning arcs from storm cloud to the ground; tides rise and fall upon the beach. It is the greatest game of them all, the Game of the World.

This game too has its rules. We call them laws of nature. Many have already been discovered, others have not. But in any event we are convinced that they exist. Everything that happens takes place according to the rules—the intricate means whereby mammals maintain their body temperature, the marvelous folding and unfolding of the double helix that is DNA and its regulation of protein synthesis within the cell, the complex dance of electrochemical activity within the brain that somehow yields thought; all these, and even that most climactic, profound event of all, the event, only dimly seen at present and inconclusively guessed, in which the first living organism came into being.

The business of science is to discover the rules whereby this game takes place. The methods, of course, are indirect.

One relies on inference, deduction, and guesswork. One searches for patterns, regularities. And more than that—one hopes to stumble upon critical moments, revealing and surprising circumstances in which the underlying order is most clearly revealed.

Here is one such. Here is a clue from which the rules of the game may possibly be inferred. A condemned man stands bound to a stake before a firing squad. A soldier steps forward, raises the rifle to his shoulder, and fires point blank. The condemned man flinches . . . but the rifle has misfired.

It is tempting to call the misfiring of the rifle a happy accident. But it was not an accident. After all, the rifle merely did what it had to do. The cocking mechanism might have developed a certain amount of play, and upon being pumped failed to insert a bullet into the firing chamber. The firing pin might have been jammed by a tiny speck of dust; the gunpowder might have been wet. In any event the gun had no choice in the matter. It did not decide of its own free will to misfire. It did what it did because of its inflexible, unbending submission to the laws of physics; and *it is these laws that contain the seeds of the condemned man's life.*

The double resonance at work in the red giant stars is a direct consequence of the laws of nuclear physics. But what if those laws were different? Whatever the explanation for the electron-proton charge equality turns out to be, it will rest upon a law of nature, and it is always possible to ask how things would be if that law were invalid. The same is true of the physical processes that determine the distribution of stars within our galaxy, and the overall expansion of the universe.

Why should the rules of the game nurture life so carefully? Nothing in all of physics explains why its fundamental principles should conform themselves so precisely to life's requirements. The laws of nature could have been laid down only in the very instant of the creation of the universe, if not before, and this occurred long before the origin of life. In that instant, how could the universe have foreseen that billions of years later, on some tiny blue-white mote of dust, living creatures would have emerged—foreseen and adjusted itself accordingly?

"Putting such questions to Nature, and wringing answers

from her is often a difficult business, as every scientist knows. Yet sometimes it is as though Nature were trying to tell us something, almost to shake us into listening." The words of George Wald.

Two more soldiers in the firing squad step forward. They raise their rifles and squeeze the triggers.

Click.

Click.

7

The Light of the World

"The wind goeth toward the south, and turneth about unto the north; it whirleth about continually, and the wind returneth again according to his circuits. All the rivers run into the sea; yet the sea is not full; unto the place from whence the rivers come, thither they return again."

Thus saith the Preacher. But don't stop there—keep going. Those clouds above my head build upward into mighty cumulonimbus, then dissolve into thin air. The wind ruffles the surface of a lake and lifts the sea into great waves; these waves perpetually carve and recarve coastlines into ever new forms. The faint babble of a running brook delights my ears, while not far away a mighty waterfall thunders. A girl sprints down the beach, kicking the sand before her and disturbing its conformation; tugboats thrust up hills of foam before them, while overhead a mighty airliner roars. In a score of nations gunfire breaks forth: battles, revolution.

The author of Ecclesiastes wrote of the futility of all things. I write of their wonder. Everywhere there is motion, flux. But why? This great and ancient globe of ours has been floating in space, entirely undisturbed, for billions of years. You'd think it would be getting a little tired by now. After all, things left to themselves usually come to rest: The swinging pendulum soon halts unless set in motion again; a person locked in a hermetically sealed room and denied food, water, and air soon dies. But the Earth, deprived of outside pushing, deprived of external food, water, and air, shows no sign of slowing its ceaseless activity. What winds up its spring?

The Sun does. The Sun is the light of the world. More than that—it is the life of the world.

If for some reason the Sun were to go out, the last, ultimate night would fall, a perpetual darkness unrelieved by any dawn. In such an eventuality the Earth would rapidly cool down to the dreadful, annihilating temperature of interstellar space, close to 460 degrees below zero Fahrenheit. Less obvious but also true is the fact that the steady bustle of the world would rapidly come to a halt. It owes its existence to the Sun. Without exception, each of the activities described above flows from it. The wind blows because differing regions of the atmosphere are heated differently; rain falls, collecting into streams and ponds, babbling in brooks and thundering in waterfalls, because sunlight has evaporated water from the ocean, tons of it each day wafted into the air later to fall as rain. The child running on the beach could not survive without the Sun's warmth, nor without the food she eats, which itself is made from living matter—and as emphasized above, all beings depend through photosynthesis upon the Sun. The tugboat chugging across the harbor, the jetliner in its flight, and the automobile puttering down the road all run on fuel derived from oil, which itself was formed ages ago by the action of geologic processes on buried organic matter; cars run on ancient sunlight. The same is true of our other fossil fuel, coal, which is made from long-dead plants. Life exists here on the Earth, but the seat of that life, its warm beating heart, is not here. It is in the Sun. It *is* the Sun.

But the Sun's existence hangs by a thread.

*   *   *

As emphasized in Chapter 3 all things, the Sun included, are made of subatomic particles: electrons, protons, and neutrons. That chapter concentrated on the electrical force, which is seated in the electron and the proton and in which the neutron takes no part. But the neutron does have a crucial role to play in maintaining the Sun's existence.

The neutron and the proton are very similar. Unlike the electron and proton, these two particles have the same size, undergo the same nuclear reactions, and participate in the strong interaction in the same way. In many ways the neutron simply seems to be a proton stripped of its electric charge. There are, though, two important differences between these particles. One is that, unlike the proton, the neutron requires other particles—either protons or other neutrons—in close proximity. Deprived of them it splits apart, decays, and it does so in about ten minutes. The neutron's protective neighbors are found in only two places: within the nuclei of atoms, and within the superdense hearts of neutron stars, which in many ways can be thought of as oversize nuclei. Only in these two locations, therefore, does that particle exist, and if one were to reach into either with a pair of tweezers and remove a single neutron, it could survive only ten minutes. After that length of time it would decay. It would decay into a proton and an electron.

The second difference is that the neutron outweighs the proton. This fact is somewhat surprising, for as explained in Chapter 11 both the neutron and proton are themselves constructed of yet smaller particles known as quarks; there seems no fundamental reason why the neutron should be the more massive of the two. Furthermore, the mass difference is quite small, a mere tenth of a percent. One might think it would make no difference. But it does make a difference. Indeed, it is crucial.

In one hand I hold a stone, in the other a hammer; whacking the hammer against the stone I break it into two parts. The splitting apart of the neutron is analogous to the splitting of the stone, and it is important to note that this stone cannot be broken into pieces *bigger than it is*. Similarly, the neutron is free to decay only because it is heavier than the proton.

Indeed, it outweighs the proton and electron combined.

But what if it did not? What if the proton outweighed the neutron?

Within both atomic nuclei and neutron stars few differences would result. But for particles in isolation the situation would be reversed. Now it would be neutrons that were stable and protons that decayed—they would decay into neutrons. And isolated protons are quite common in nature: They are found in hydrogen. The atomic nucleus of hydrogen consists of nothing more or less than a single proton. That element therefore depends for its existence on the stability of the proton, and this in turn depends on that microscopic difference in mass between the proton and the neutron. If the difference were reversed, hydrogen would not exist.

Chapter 4 discussed the properties of water, and their relevance to the requirements of life. That discussion left something out, though: The stuff has to *exist*. But water is H_2O. If hydrogen did not exist, water would not exist.

Speculations have occasionally been made that some form of silicon-based life form might be possible. It is an open question whether these speculations are to be taken seriously, and if so whether such beings could get along without water. Let us grant both for the sake of argument. Even so, however, the absence of hydrogen would prove fatal to such hypothetical beings, for whatever the details of their biochemical organization, they still would require an outside source of energy—a sun. *But the Sun is made of hydrogen.*

If at this very moment some mysterious process were to increase the proton mass relative to that of the neutron, little would happen for several minutes. But after that interval of time the protons in the Sun would commence decaying. Soon the Sun would be made not of hydrogen but of neutrons. And what would such an object be like?

The central difference between such a beast and the true Sun lies in the insulating properties of the material of which they are composed. In this regard the Sun is much like a house. In constructing a house it pays to install good insulation, for if one does not the building is unprotected against the cold of winter. The Sun, likewise, floats in the absolute zero of space, and the heat it loses must be replenished. But that re-heating takes energy, fuel; and there is only so much of it available to the star. Ultimately the Sun is going to run out of fuel. The poorer the insulation, the more rapidly will this occur . . . and the more rapidly will the light of the world go out.

Neutrons, it develops, make terrible insulation—hundreds of millions of times poorer than the material of which the Sun is actually composed. If the protons within it were to decay, the Sun would commence burning fuel at a rate so enormous that within a century it would be entirely exhausted. And after that time the Sun would go out.

Nor could the Sun have survived more than a century after its formation. But one hundred years is not long enough for life to develop, let alone for evolution to proceed. In biological terms it is the merest flicker of an eyelash. Furthermore the Sun is not alone in this regard, for most stars in the sky are composed of hydrogen. None could have existed beyond that microscopic interval of time had the proton outweighed the neutron. Of course certain other stars, the red giants such as Capella, Pollux, and Aldebaran, are composed not of hydrogen but of helium. They would hardly be affected in such a circumstance. But while red giants play a vital role in preparing the cosmos for life they do not make suitable seats for it, and for the same reason: They do not survive long enough. The same is true of every other stellar type. Only stars like the Sun, those composed of hydrogen, maintain themselves for the vast intervals of time required for the full development of life about them; and these in turn exist only by virtue of that tiny offset—one tenth of a percent—between the masses of two subatomic particles.

That is another misfired rifle in the firing squad.

Here's another.

In Chapter 4 the process of photosynthesis was discussed, and its importance to life. That discussion concentrated on the role of the H_2O molecule in the photosynthetic reaction. It omitted to mention, however, a second factor: a quite extraordinary matching between the properties of the Sun and those of plants. Failing that match, photosynthesis could not take place.

Chlorophyll is the molecule that accomplishes photosynthesis. This molecule is universal in plants, and it is what gives leaves, grasses, and the like their distinctive green color. The mechanism of photosynthesis is initiated by the absorption of sunlight by a chlorophyll molecule. But in order for this to occur, the light must be of the right color. Light of the wrong color won't do the trick.

A good analogy is that of a television set. In order for the set to receive a given channel it must be tuned to that channel; tune it differently and the reception will not occur. It is the same with photosynthesis, the Sun functioning as the transmitter in the analogy and the chlorophyll molecule as the receiving TV set. If the molecule and the Sun are not tuned to each other—tuned in the sense of color—photosynthesis will not occur.

As it turns out, the Sun's color is just right. Light from the Sun is readily absorbed by chlorophyll, so initiating the photosynthetic process. On the other hand, the Sun's color is related to its temperature: Things heated moderately glow a dull red, but if heated yet more glow a brilliant yellow. Thus the matching under consideration is between the temperature of the Sun on the one hand and the molecular structure of chlorophyll on the other. Without that matching, life could not exist upon the Earth.

Nor could life exist anywhere else. It is not a matter of seeking out the correct niche—some star of the right temperature. As emphasized above, only stars made of hydrogen are suitable seats of life, and it turns out that all such stars have roughly similar colors. From the coolest to the hottest, the variation is not great. Thus starlight impinging upon all planets, wherever they may be in the universe, has a roughly similar color. Either all stars provide good niches or none of them do.

One might think that a certain adaptation has been at work here: the adaptation of plant life to the properties of sunlight. After all, if the Sun were a different temperature could not some other molecule, tuned to absorb light of a different color, take the place of chlorophyll? Remarkably enough the answer is no, for within broad limits all molecules absorb light of similar colors. The absorption of light is accomplished by the excitation of electrons in molecules to higher energy states, and the general scale of energy required to do this is the same no matter what molecule you are discussing. Furthermore, light is composed of photons, packets of energy, and photons of the wrong energy simply cannot be absorbed. The energy carried by these photons, though, depends upon the color of the light. In terms of the above analogy, it is simply not possible to build a TV set capable of receiving signals differing wildly in frequency from the normal. If some station, more enterprising than wise, elected to transmit at such a radically differing frequency, no sets would be capable of receiving its signal. It would lie outside their range.

Similarly, if stars were slightly different in their temperatures no great differences would arise. But the light from radically cooler stars could not be absorbed by molecules of any sort. Photosynthesis could not occur—not the photosynthesis we know, employing chlorophyll, nor any conceivable form of it employing electron excitation in molecules of any description. A planet huddled up close to such a hypothetical star, say the distance of Mercury from our sun, could easily be warm enough to support life, and it might well possess large amounts of water. That planet would have oceans, rain clouds, warm and balmy breezes . . . but no plants, and therefore no animals. Alternatively, if stars were far hotter their emission would pack such a wallop as literally to tear molecules apart, dissociate them. A high-temperature star would emit not visible light but ultraviolet, known to be dangerous; still hotter ones would emit X rays. Planets far from such stars, say the distance of Pluto from our sun, could hold oceans, rain clouds, warm and balmy breezes . . . but down upon their ravaged surfaces would pour a flood of sterilizing radiation. As things stand in reality, there is a good fit between the physics of stars and that of molecules. Failing this fit, however, life would have been impossible.

*   *   *

THE LIGHT OF THE WORLD

Some more weapons in the firing squad.

The Sun is a furnace, burning fuel; the energy liberated by that burning is the source of its great light and heat, the beating heart upon which life on the Earth depends. If the furnace were to go out all life would cease. But there are some potential problems associated with that fuel and its utilization by the Sun.

The fuel is hydrogen, the burning its transmutation by nuclear reactions into helium. Just as the red giant stars transmute helium into carbon, so do the Sun and stars like it prepare the stage by first producing that helium. The ashes from the operation of one furnace become the fuel used by the next. And because the furnace burns nuclear fuel, in essence solar energy is nuclear energy. The only difference is that in this case the reactor is 93 million miles away.

As discussed in Chapter 1, the helium nucleus consists of two neutrons and two protons, as in Figure 12, while the nucleus of hydrogen is a single proton. The reaction powering the shining of the Sun is the amalgamation of four nuclei of hydrogen to form one of helium; in the process two of the protons turn into neutrons (Figure 13).

Helium: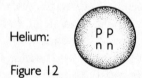

Figure 12

Just as in the red giants, such a multiple encounter—a four-way collision in this case, among particles flying about randomly—is an exceedingly unlikely event. Therefore the formation of helium from hydrogen does not occur in this direct fashion. Rather, once again as in the red giants, it happens in stages: First two protons combine to form some intermediate nucleus, and then this nucleus combines still further to yield helium. The intermediate nucleus so produced is the deuteron, consisting of one proton and one neutron. The first stage in the burning of the sun's nuclear fuel is therefore as diagramed in Figure 14.

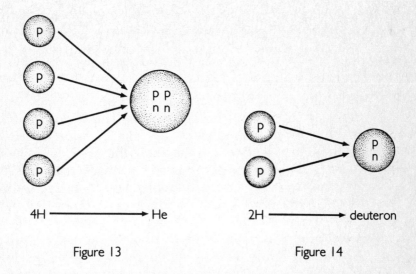

4H ────────────► He

Figure 13

2H ────────────► deuteron

Figure 14

What holds the deuteron together? What glue is responsible for the sticking together of a proton and neutron to form that nucleus—or, for that matter, of any number of protons and neutrons to form any atomic nucleus? There is a force that accomplishes this; a force that, in addition to binding together the nucleus, fuels the nuclear reactor as well as the atom and the hydrogen bombs, and powers the shining of the stars. It is known as the strong force.

The strong force is the most powerful force known, very much stronger than even that of electricity. Thus thermonuclear weapons, which employ it, are so much more dangerous than ordinary ones, which do not. Nevertheless it develops that this force is not really so very powerful. In fact, it is only barely strong enough.

It is only barely strong enough to bind together the proton and neutron to form the deuteron. That nucleus, in fact, is only marginally bound. Were the strong force slightly weaker—about half as strong or less—the deuteron would

THE LIGHT OF THE WORLD

simply not exist. Two protons brought together in an attempt to form it would merely bounce off one another and fly back away into space. And in such a circumstance that intermediate nucleus, so crucial to the formation of helium, would be absent.

In that case, the nuclear reactions powering the shining of the Sun could not proceed. Nor is there any resonance at work here to circumvent the bottleneck. The lucky double matching at work in the red giant stars simply does not exist in this situation. Deprived of its fuel, the Sun could not shine. Most other stars in the sky, also composed of hydrogen and employing the very same reactions as the Sun, also could not shine. These stars, in turn, produce the helium that red giant stars employ as fuel . . . so red giants would not exist, nor indeed any other stellar type. And so, had the strong force been only slightly less strong, the light of the world, the lights of all the worlds, would never have been lit.

What if that force had been stronger?

A glance at the reaction illustrated in Figure 14 shows that two quite different things are going on within it. On the one hand, a proton is turning into a neutron. On the other, the neutron so produced combines with another proton to make the deuteron. The second of these processes involves the strong force. But concentrate now upon the first.

The transformation of a proton into a neutron only occurs in the course of nuclear reactions: The proton undisturbed and in isolation is stable. On the other hand, that transformation is obviously related to the inverse process, the decay of the neutron. That decay takes ten minutes, and while this period of time seems fairly short in human terms it is not so in the context of subatomic processes. Just the opposite: It is enormously long. The slowness of this decay signals that the force responsible for it cannot be the same as that responsible for the gluing together of nuclei. If the strong force governed such processes they would be very fast. Thus an entirely new phenomenon is at work here, the so-called weak force.

The formation of the deuteron therefore involves two separate forces, the weak and the strong. It is a chain . . . and a chain is only as strong as its weakest link. The burning of fuel within the Sun proceeds through a bottleneck, the weakness of the weak force, and as a result it proceeds only with difficulty. Just as the neutron takes a long time to decay, so the transformation of hydrogen into helium goes slowly. And because it does so, the fuel upon which the Sun operates is not very good.

But if the strong force were stronger the picture would change entirely. In that circumstance an entirely new nucleus would enter the picture: a nucleus composed of two protons bound together (Figure 15). This object, known as the di-proton, does not exist in reality. If one tries to create it, the thing immediately flies apart. On the other hand, it only barely does not exist. Were the strong force stronger—about twice as strong—the di-proton would come into being as a stable nucleus . . . and the Sun would change beyond recognition.

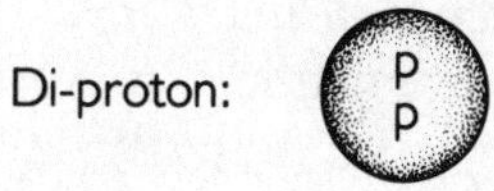

Di-proton:

Figure 15

The Sun would change because the first stage in the formation of helium would no longer be the formation of the deuteron. It would be the formation of the di-proton, as in Figure 16. And this reaction would not involve the transformation of a proton into a neutron at all. The role of the weak force would be eliminated, and only the strong force would be involved. The weak link in the chain would be removed, and as a result the Sun's fuel would suddenly become very good indeed. It would become so powerful, so ferociously reactive, that the Sun and every other star like it would instantaneously explode.

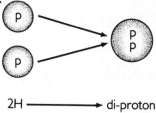

2H ⟶ di-proton

Figure 16

THE LIGHT OF THE WORLD

That is what would happen were the strong force stronger. Or at least that is what many scientists think would happen. For years now this particular argument has been advanced as yet another example of a misfired rifle in the firing squad. But I myself am not so sure.

I am not so sure because the Sun is not just a furnace burning fuel. It is a *thermostatically controlled* furnace. The thermostat here is the size of the Sun. Were it to commence burning fuel at an enhanced rate, the Sun would grow hotter. On the other hand, in response to this increased temperature it would then grow larger, puff up like a balloon—but expanding a gas cools it down, and so reduces the rate of burning of that fuel. In this way the Sun, and every star like it, maintains a fine control. Stars are not bombs.

Rough calculations confirm that were the strong force stronger, the Sun, rather than burning fuel at a monstrous rate, would indeed grow larger and cooler—much larger and much cooler. On the other hand, my own feeling is that these calculations are not yet sufficiently advanced to provide the definitive answer to this question. Among other things, they do not address the question of the *stability* of burning of such ferociously reactive fuel. The stuff may turn out to be too dangerous, too prone to violent flare-ups. A further question is whether the color of the light emitted by such very cool stars would still fit the absorption properties of chlorophyll— or indeed of any other molecule.

I am confident of the validity of the other arguments advanced in this chapter. But until more work is done on the last, it is too early to say. On the one hand, it might turn out to be a genuine misfired rifle. On the other, it may simply be a dud.

8

Space

Fix your attention on some object. What object? No matter—anything will do. It could be a chair, a table, the furnace in the basement, or a shingle on the roof. All I want to know is where it is. My question is, in what direction does that particular thing lie from you.

To answer that question, three separate responses must be given. You must specify whether the object lies in front of you as opposed to behind, to the right as opposed to the left, and above as opposed to below. Any briefer answer would be incomplete. Directions, of course, are directions in space; to say that the specification of direction involves the specification of three different things is to say that space has three dimensions.

So accustomed are we to this fact that we hardly ever reflect upon it. But we should, for it is intimately related to our existence. The British astronomer G. J. Whitrow pointed out in 1955 that if space did not have three dimensions, we would never have come into being.

Begin by considering an imaginary world of two dimensions, a universe existing on a plane. That is the world figures drawn on a page inhabit. In it, the dimension forward-and-back exists, as does right-and-left. But the third dimension, up-and-down, has vanished without a trace. It turns out that we could not function were reality constructed in such a way. In particular, it would be impossible to think in such a universe.

The problem is one of scrambling, and it can be illustrated with the analogy of a telephone network. If space were two-dimensional, every telephone in the world would be connected to a multitude of others. Each time I picked up the phone I would find myself struggling against an immense gabble in my effort to make myself understood: "Thank you for calling Delta All our lines Anyway I'll be checking in with Milly After two hours of that crap I told that idiot he could She can't just sit there on it I mean how long The number you have reached . . ."

And so forth. Against such an uproar no communication would be possible. That is what would happen to the telephone system if for some reason space were suddenly to flip from three dimensions to two. More than that, it is what would happen to our brains. The operation of the nervous system, upon which mental processes depend, would become impossible.

The brain is composed of brain cells, or neurons. From each neuron a number of delicate fibers, known as dendrites and axons, extend out in all directions. Each fiber leads to another neuron, and rests upon it. Through these fibers pass nerve impulses: first into the dendrites, from them next into the main cell body, and finally out along the axon. In this manner, via dendrites and axons, does the nervous system's internal communication take place. The configuration is diagramed in Figure 17.

The analogy with the telephone system is a good one. In the analogy, each neuron is represented by a household.

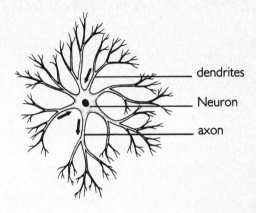

Figure 17

Imagine a network in which not one but many separate telephone lines led out from each home, each linked to a different home, and each open and operating constantly. Sitting
in my room I would be incessantly in communication with
others in every corner of the world—one in Arizona, another
in Paris. Over some lines I would listen, over others I would
speak; never pausing, never hanging up.

That is a reasonable model of the workings of the brain—
and not just of human brains, but those of all higher animals.
No one pretends to understand fully how thought, awareness,
and perception arise from this perpetual cross-pollination.
Nevertheless, three general features seem clear. The first is
that *many* neurons are required for thought. The human
brain contains a truly remarkable number of cells: tens to
hundreds of billions; in terms of the analogy, that is far more
than the number of homes in the world. The same is true of
the brains of the other higher animals. At the other end of the
spectrum lie those organisms with rudimentary brains, which
invariably turn out to exhibit only the most primitive forms
of behavior. An example is the snail, the nervous system of
which involves a mere few thousand neurons. No one ever
held an engrossing conversation with one of those beasts, nor
found their play a source of fascination and delight.

Second, it is clear that *intercommunication among the
neurons* is required for thought. The neurons themselves are

not enough—they need to maintain that perpetual exchange via their dendrites and axons. The exchange is marvelously complex, for neurons can possess enormous numbers of separate dendrites; at each moment, every cell in the human brain is in communication with from hundreds to tens of thousands of others. But though complicated, the workings of the brain are not random, and that is the third and final point: Each fiber of the network leads to a definite recipient. This *orderly contact* is crucial to proper brain function. Only if my phone lines lead to Arizona and Paris will things go correctly; if the system is scrambled and I suddenly find myself speaking to Detroit and Tokyo, its operation will fall apart.

But that order is not possible in two dimensions.

Go back to the telephone system. To preserve the analogy with brain structure, make sure the network relies exclusively on telephone lines—analogues of dendrites and axons—strung from household to household, rather than communication via satellite or microwave link. Build it up in stages, beginning with one line, from Boston to Phoenix, say, as in Figure 18. Now add a second, this one from Chicago to St. Louis (Figure 19).

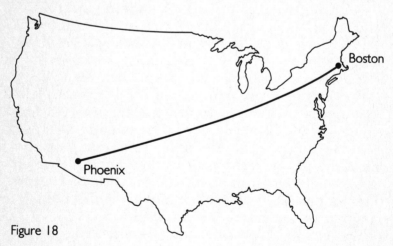

Figure 18

Those two lines cross in Kansas. It looks from the diagram as if they touch there. They do not, of course; the phone company suspends them at different heights on the telephone

Figure 19

poles. They might be a mere few inches apart where they cross, or a comfortable several yards; in any event, they are not making contact.

But they would make contact if space had two dimensions. The third dimension is up, along the telephone pole and into the sky—and the phone company makes use of it in stringing the lines. If "up" is not available they will touch where they cross, just as the two lines of ink representing them blend into each other and merge in the two-dimensional Figure 19. In such a circumstance the Boston-Phoenix link would be amalgamated with that between Chicago and St. Louis. And that is the difficulty facing brain function in two dimensions: Communication becomes scrambled. Each dendrite and axon leading to its intended neuron would make unwanted contact with others along the way.

But surely there must be some means of stringing those lines to avoid the difficulty! Suppose one of the two is re-routed, as in Figure 20. It will not do, for now the Chicago–St. Louis line interferes with the one from Phoenix to Los Angeles (Figure 21). And if the line is routed to the east instead, it runs into other difficulties—communication with Europe, for instance. No matter how many possibilities one tries, nothing works.

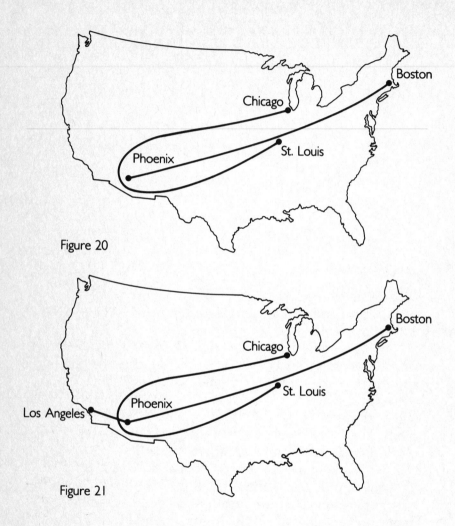

Figure 20

Figure 21

There are of course more lines in the network, for example that between Minneapolis and Atlanta (Figure 22). Now there are more unwanted connections. The more cities one adds, the worse the situation grows; also, the more phones in each city, the worse yet again. Everyone speaks with unwanted partners. And because the brain contains so many neurons, the confusion is immense.

There is an example known to medical science of the results of such a disorder of the nervous system. Just as tele-

Figure 22

phone wires are insulated to prevent their signals from leaking into one another, so axons are insulated for the very same purpose by a substance known as myelin. A disease is known in which this myelin degenerates. The insulation is stripped away, and each neuron's channels of communication become mingled with those of others.

The name of that disease is multiple sclerosis.

The conclusion is that you need three dimensions to think. As a matter of fact, you also need three dimensions to live, for the difficulty is not confined to the nervous system; any biological function involving flow along some well-defined pathway becomes scrambled in this manner. A second example is that of blood flow. In two dimensions arteries would intermingle with veins, blood from the heart destined for the hands would wind up in the kidney, and oxygen-poor blood, rather than passing to the lungs, would somehow reach the brain. No creature could survive such a scrambling.

Dropping down from two dimensions to one, the situation becomes yet more horrible. Moving up to three dimensions, on the other hand, solves the problem. Continuing the progression, one is naturally led to consider a fourth dimension.

Relativity has demonstrated that time can be regarded as a fourth dimension, but that is not the subject here; the question is whether a fourth direction might exist, something lying beyond the usual up-and-down, forward-and-back, and right-and-left. The notion has exercised a perpetual fascination and has led to endless conundrums beloved of science fiction writers. On the other hand it seems inconceivable. We are, in fact, quite incapable of visualizing such a thing. No one knows how to imagine it. But while such a concept cannot be visualized, it is easy to analyze mathematically. And by using such mathematics it is possible to show that, had space more dimensions than three, complex networks could be interconnected yet more efficiently and reliably. You can think better in four dimensions than three.

On the other hand you cannot *live* better in four dimensions than three. Indeed, you cannot live at all. In four dimensions the Earth's stable, life-supporting orbit about the Sun could not exist.

As I write these lines a soft breeze blows through the open window. It is a perfect summer day, not too hot, not too cold. Come to think of it, it was also neither too hot nor too cold last January, when the thermometer hit 20 below—nor last week, when we sweltered under a 90-degree heat wave.

It's all a matter of expanding one's point of view. The lowest possible temperature is absolute zero, 460 degrees below zero Fahrenheit. That is the temperature of interstellar space, very nearly. The Sun's surface provides a second reference point: It is close to 10,000 degrees Fahrenheit. Those are the limits between which the Earth's temperature is potentially capable of swinging. And compared to that range, the actual swing from the heat of summer to the cold of winter is minuscule.

All in all we live in a benign environment, conducive to life, and remarkably so considering the range of possibilities: a sheltered, protected corner of the cosmos. It has to do with being in the right place—in the right niche. Were the Earth much closer to the Sun, the heat would be insupportable, while too far away everything would freeze. As it is, our location is exactly right.

But as before, the real question is not why we live in this good niche. It is why the niche exists at all. It is not enough for the Earth simply to be located at the appropriate distance from the Sun—it must stay there. It must not venture too far from its present location if life is to be possible upon it. In terms of orbits, this means our planet's path about the Sun must be circular. As discussed in the Prologue, were it not circular the immense fluctuations in temperature as we alternately approached and receded from that warming fire would be deadly.

There are, though, a number of things that might disturb our nicely circular orbit. One is the close passage of a wandering star. Also as emphasized in the Prologue, this is a highly unlikely event in view of the enormous scale of the astronomical universe. We are safe in that regard. But could our orbit be disturbed in some other way?

Imagine a marble and a cone—a dunce cap. The marble is perched atop the cone. A precarious balance; the slightest tremor, the most infinitesimal of disturbances, and that marble will topple sideways to the floor.

In this example there is no need to ask what external agency might disturb the configuration. *Anything* will disturb it. The marble's balance is fragile, desperately so. In technical terms we say the balance is unstable. The question is whether the Earth's orbit could be similarly unstable.

Clearly it could not—for if it were, something would have disturbed the orbit millennia ago. The smallest of meteors would have struck the Earth, giving the planet the slightest of punches; Mars would have swung from one point in its orbit to another, so changing the direction of its gravitational attraction; a faint puff of solar wind would have shoved us slightly outward . . . and in response the Earth would have slid from its present location, plummeting into the Sun or rocketing off into interstellar space. Either eventuality would have been a disaster; we would have burned or frozen.

If orbits were unstable, no celestial body could remain in orbit: not the Earth, not any of the other planets, not even distant worlds swinging about their own suns. Rather, every planet in the universe would wander freely through the icy void of interstellar space, bound to no warming star—or, alternatively, would plummet into a star. Life could not exist in such a circumstance. Niches would not exist.

Obviously this does not occur. Orbits are stable. Rather

than being like a marble on a cone, they are like a marble in a teacup. But *why* are orbits stable? They are stable because the fourth dimension does not exist.

The way to decide whether a configuration is stable or unstable is to disturb it slightly and see what happens. Consider first the stable configuration of a marble in a teacup. If that marble is moved slightly away from its home location in the cup's center, the trend of events is for the marble to return (Figure 23). That is the mark of stability. On the other hand, a marble on the tip of a cone, if moved away from that tip, tends to move yet farther away (Figure 24). That is the mark of instability.

Figure 23   Figure 24

Do the same with the Earth in its orbit. Consider the effects upon that orbit of the tiniest of meteors burning up high above our heads as a shooting star. That meteor has pushed against the planet, however slightly, and depending on its direction of motion the push is either toward or away from the Sun. For the sake of concreteness, consider an inward shove. In response, the Earth drifts microscopically inward—but then it quickly moves outward, back to where it began before the shove (Figure 25). But why is the motion back outward?

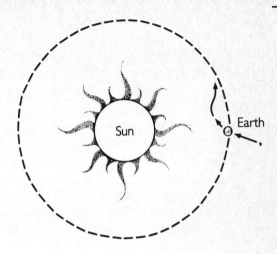

Figure 25

Prior to the meteor's shove the Earth, orbiting the Sun, is moving in a circle. A force is required to bend the path of our motion into this circle: Lacking this force the Earth would not orbit the Sun at all. Furthermore it turns out that the smaller the circle—the closer in we lie—the stronger is the force required. What is this force? It is the gravitational tug the Sun exerts upon the Earth. Orbits are made by gravity.

Once the meteor shoves the Earth toward the Sun, however, were the planet to remain in this new location—or worse, move yet closer—we would be in trouble, for the orbit would be unstable. The marble would have begun its roll down the cone. On the other hand, this would necessitate forcing the Earth's motion into a tighter arc. And that's hard to do; a greater force of attraction is required. Newton's famous inverse-square law of gravitation holds that the more closely the Earth approaches the Sun, the more strongly the Sun attracts it. *The test of stability is the rapidity with which this occurs.* Were the force of attraction to increase very rapidly, our orbit would be unstable.

In reality, though, the rate of increase is mild. The Earth's orbit is stable. But the rapidity of increase, in turn, is related to the dimensionality of space.

*   *   *

The way to think of gravity is in terms of lines of force, radiating outward from the Sun like spines on a sea urchin. The more tightly packed these lines, the more powerful the gravitational pull. Why does the degree of packing depend on the dimensionality of space? Begin with a hypothetical universe consisting of merely one dimension. In such a world everything lies on a line, and the only directions that exist are forward and back. The Sun's lines of force are bundled together into a sheath, as in Figure 26. They lie evenly together, growing neither more nor less concentrated as they extend outward from the Sun. Thus, had space one dimension, gravity would grow no stronger as the Earth approached the Sun. That is the hallmark of *stability*.

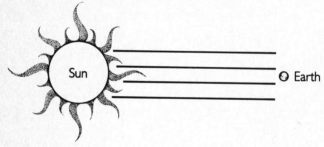

Figure 26

The two-dimensional universe of Figure 27 is different. Here the packing close in is tight, farther out less so. Gravity grows stronger as the Earth approaches the Sun. But it does not do so very rapidly—not sufficiently rapidly to produce an unstable orbit. This too shows the hallmark of *stability*.

Now the real, three-dimensional universe. Here the diagram cannot be drawn upon a flat page—we need a model. Such a model might be a sea urchin with its spines; alternatively, one can be built by sticking wires into a tennis ball. By fiddling with such models it is easy enough to see that lines of force draw apart from one another more rapidly in three dimensions than in two. As the Earth approaches the Sun, gravity grows stronger more rapidly in the real, three-dimensional universe than it would if space had only two dimensions. But it still does not increase sufficiently rapidly

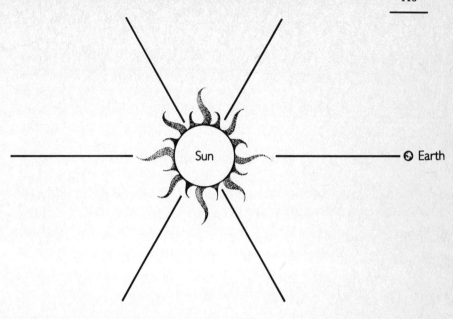

Figure 27

to produce an unstable orbit. Here too, the hallmark is of *stability*.

By now the progression is clear. Every time a dimension is added to the analysis, we edge closer toward the unstable situation. Three dimensions turn out to be the limit—and four lie beyond that limit. A model cannot be built to analyze the four-dimensional case, but the mathematics is easy enough. *If space had four dimensions, orbits would be unstable.*

It is a remarkable situation. How many dimensions of space are possible? So conditioned are we to three that entertaining the notion of four seems revolutionary. But be brave: Why not five dimensions—or five hundred? Surprisingly enough, there is nothing logically impossible about either. Mathematicians deal with them every day. There are, in fact, an infinite number of conceivable dimensions. But the above argument concerning planetary orbits can easily be extended to yet higher dimensions, and it yields the same conclusion.

Of the entire range of conceivable dimensions, only one number—three—is amenable to life. Any choice above three makes it impossible for planets to remain at proper distances from their suns. Anything below three scrambles the orderly communication so crucial to living beings. And reality chose the ideal number.

How did the universe do it? Physicists have not the slightest idea. No theory known to science comes close to explaining the mystery of space itself. Is space even subject to natural laws? Or is it an accident? If so, if reality chose the number three randomly, pulled it out of a hat, so to speak, it did so from an infinite range of possibilities. And the odds of doing that successfully are zero.

9

The Watchmaker

A man clad in archaic garments is striding upon a heath. For some reason, whenever I attempt to visualize him the image comes to mind of a portly gentleman, white-haired and florid, and of a serious mien. His name is William Paley.

A glint from something bright and shiny catches Paley's eye. He bends down and picks it up. Unlike everything else in the scene about him (soil, heather), the thing is polished to a smooth perfection. Studying it more carefully Paley finds that its face, protected from the elements by a circle of glass, is clean and white, and upon it the numbers 1 through 12 are painted. Also upon the face are two thin arrow-shaped bars, one pointing to the numeral 3, the other up to 12.

The other side has hinges on it. Prying it open, Paley finds inside an extraordinary profusion of gears, springs, and levers, all in motion. Incredible mechanism! Glancing again at the front, he finds the longer bar has moved somewhat. Eventually he comes to the conclusion that the object he has found keeps time.

But that is not what intrigued William Paley so much in the year 1802. He did not want to know what this thing he had found upon the heath did. What he wanted to know was how it had come into being in the first place. And from the intricacy of its construction, and the clear evidence of purpose in its design, he came to an important conclusion: Someone must have made it. From the design of a watch, Paley inferred the existence of a watchmaker.

Now something else attracts Paley's eye: a bright and pretty flower sprouting from a bush. He bends down to examine it. This too repays close examination. More intricacy and profusion of design—pistils, stamens, and petals with delicate shadings of color! Transferring attention to the bush as a whole, Paley finds its structure equally complex. Moreover, upon exhaustive study he realizes that, like each part of the watch, each element in the structure has a function. The bush's roots provide two services: They grip the soil and so provide support, and they suck nutrients from that soil. Within the branches a delicate network of pathways transports those nutrients about the organism. As for the flower, it is a reproductive device: Its bright color distinguishes it from the otherwise drab environment and so attracts bees, which upon penetrating the flower are brushed with pollen. The bees then transfer this pollen to a second flower elsewhere, so effecting cross-pollination.

And from the intricacy of construction of the bush, and the clear evidence of purpose in its design, Paley infers something else—not a watchmaker, but a Watchmaker.

Paley lived in an age of explosive scientific growth. Naturalists scoured the Earth, discovering and describing in exhaustive detail new species; telescopes revealed the grace and delicacy of Saturn's rings, the immense and perfect symmetry of the solar system; anatomists dissected flowers, eyes, hands, wings, jaws, beaks, and, in general, everything they could get their hands upon. His was a time in which the full intricacy of the natural world was beginning to be appreciated. And more than intricacy—there was design as well.

From its existence Paley drew an important conclusion. "The marks of design are too strong to be gotten over," he wrote. "Design must have a designer. That designer must have been a person. That person is GOD."

Natural Theology was the title of his book, and it is an intriguing combination of words (the *natural* is used in the sense of *naturalist*). Paley, inventor of the famous argument of the watch upon the heath, was a doctor of divinity and archdeacon of Carlisle, but the book he wrote was more crammed with recent scientific discoveries than with religious exhortation. He was using science to demonstrate the truth of the Christian faith.

So too was William Whewell, master of Trinity College and professor of moral philosophy at the University of Cambridge. Whewell was a scientist and the author of a widely used textbook on calculus—and a D.D. as well. Like Paley, his view of the use of science was ultimately religious in nature. He wrote of "a harmonising, a preserving, a continuing, an intending mind; of a Wisdom, Power and Goodness far exceeding the limits of our thoughts." Every fresh discovery, he wrote, "suggests to most minds the belief of a creating and presiding Intelligence."

Astronomy and General Physics Considered with Reference to Natural Theology was the title of a book by Whewell. It was published in 1833 as one of a series of so-called Bridgewater Treatises, endowed from the bequest of the late Reverend Francis Henry Egerton, earl of Bridgewater, who, when he died four years earlier, had left the sum of eight thousand pounds sterling to support publications demonstrating "the Power, Wisdom, and Goodness of God, as manifested in the Creation." The operative word here is "manifested." Bridgewater had no use for religious fanatics. He wanted scientific arguments, strict, rational proof of the Lord's existence, and the eight treatises published under his bequest ranged in subject matter from anatomy to zoology, from meteorology to chemistry to physics and astronomy. Everything was grist for the Bridgewater Treatises' mill.

Paley wrote of the human eye, and he showed how similar it was in construction to the astronomical telescope. Like the telescope, the eye possessed a lens to gather light and focus it, and a means of adjusting this focus to differing distances. Remarkably enough, Paley noted, the eye also possessed an ingenious means of correcting for chromatic aberration, the unequal focusing of light of differing colors. This was a problem that lens grinders had unsuccessfully struggled to overcome for years. Ultimately, someone conceived the idea of asking how nature did it. Adapting what he found in the human eye to the astronomical telescope, this man succeeded where others had failed. "The most secret laws of optics must have been known to the author of a structure endowed with such a capacity," Paley commented.

He wrote too of the mechanical arrangement of the skeleton. The head, for instance, had to be capable both of nodding forward and backward and of turning from side to side. In response to this need, the neck was provided with two quite separate mechanisms. On the one hand there was a hinge allowing the head to nod; on the other

> what anatomists call a process, namely,
> a projection somewhat similar in size
> and shape to a tooth; which tooth
> entering a corresponding hole or socket
> in the bone above it, forms a pivot or
> axle, upon which that upper bone,
> together with the head which it
> supports, turns freely. . . .

Discussion is to be found in his book of the knee, elbow, and wrist; of the arrangements of muscles and of the flow of blood.

In his Bridgewater Treatise, William Whewell wrote of the remarkable appropriateness of the length of the year to the needs of plants. They turned out to possess cycles of activity lasting precisely this period of time. Each spring the maple put forth leaves; six months later these leaves turned a brilliant red and fell to the ground. The snowdrop flower thrust forth in February, the apple ripened in the fall. All these organisms contained timekeeping mechanisms—biological clocks, we call them today—and the striking thing about these clocks was that they kept the same time as the passage of the seasons.

Had this not been true, the leaf that thrust forth one year in spring would have sprouted the next year in midwinter. Having incautiously done so, it would have promptly frozen. The apple ripening too late, the snowdrop blossoming too soon—all would wither away and die. In the absence of a fit between these two clocks, the one biological and the other astronomical, the year would become "a year of confusion," Whewell wrote. It would become "a year of death."

But he knew that the length of the astronomical clock was determined by circumstances entirely unrelated to the needs of plants. It had to do with our location in the solar system. Had Earth been closer to the Sun the year would have been shorter, if farther away longer. How then had this matching between two such wildly unrelated clocks come about? In a remarkable passage Whewell made the following suggestion. He proposed that originally some plants had been created whose clocks ticked out a year lasting not twelve but eleven months, while others had been created whose clocks measured the correct year of twelve months, yet others thirteen, and so forth. If so, those plants ill suited to the length of the actual year would never have survived. After a brief interval of destruction only the twelve-month biological clocks would have remained in existence.

That is a perceptive suggestion, and one that by virtue of our present knowledge we recognize to be correct. Whewell was on the verge of discovering the principle of natural selection. But at this crucial juncture he turned back. Having made his suggestion, he immediately rejected it. He rejected it because it did not explain how the design so apparent in the construction of plants had come about in the first place. Whewell's question was the same as Paley's: Who made the clocks? And because he did not know any answer to that question other than God Himself, he was prevented from making one of the greatest discoveries of all time: the theory of evolution.

Evolution has four parts, and only one of them was in Whewell's hands. The first part is mutation, random alterations in the genetic code. The second part is inheritance, the passing down of these alterations from generation to generation. The third is Whewell's natural selection, the weeding out of those mutations unfit for survival. And the fourth part is time—lots of time, incomparably more of it than the proponents of natural theology dreamed possible: the millions upon millions of years required by evolution with its microscopically slow pace to do its work. Lacking the other three pieces in the puzzle, Whewell could hardly have done more than he did.

It is evolutionarily advantageous for organisms to adjust their structure in synchronism with the passage of the seasons, in order to respond to alterations both in temperature and in the patterns of activity of other organisms. Thus it is that evolution made biological clocks. It is also advantageous for an animal like *Homo sapiens* to be able to swing its head in as many directions as possible, and to employ sunlight as a means of finding prey and avoiding predators. So evolution provided for the hinge and pivot of the neck, and the optical perfection of the eye.

By now natural theology has become a discarded relic of a bygone age. None of its arguments has the slightest shred of scientific support. Its exponents were wrong, for the flower was not made by a Watchmaker after all. It was made by a watchmaker, a natural process capable of being understood. For my purposes in this book, the Bridgewater Treatises are important for one reason only: They have given the mystery of the fitness of the environment a bad name.

Many people have sensed a strong similarity between the arguments of natural theology and those of the sort that anthropic investigations have revealed. Everywhere they looked in the biological world the authors of the Bridgewater Treatises found design, a design without which life would not have been possible. Modern physics has found that a similar life-giving design exists in the structure of the physical world. Are we not, then, merely updating a point of view by now totally discredited by the advance of knowledge? Are not our arguments at heart the same as those of natural theology? Does not the theory of evolution account for all these matchings, fittings, and lucky breaks?

I would argue that it does not. Natural theology is one thing, the fitness of the environment another. The theory of evolution has shown the first to be utterly fallacious. But it has nothing whatever to do with the second.

Whewell's discussion of the matching between biological clocks and the length of the year might strike the reader as being suspiciously similar to Chapter 7's discussion of photosynthesis, and of the matching between the absorptive properties of chlorophyll and the color of the Sun upon which it depends. The theory of evolution explains how biological clocks arose and came to be attuned to the passage of the seasons. Why can't it similarly explain this matching? I claim that evolution is perfectly capable of explaining how molecules arose that are capable of capturing light—but this is all it can explain. As emphasized in Chapter 7, the absorption of light by molecules is accomplished by exciting electrons within them to higher energy states, and the general magnitude of the energy required to do this is the same no matter what molecule is doing the absorbing. Furthermore, sunlight is composed of photons, packets whose energy depends upon the color of the Sun, and photons of the wrong energy simply cannot be absorbed. As things stand there is a good fit between the color of the Sun and the energy of excitation of molecules. Failing this fit, though, no molecular receptor of sunlight of any sort whatever would have been possible. Mutation, inheritance, and natural selection could never have surmounted this obstacle had the environment not cooperated so beautifully.

Evolution cannot work in opposition to the laws of physics. Had space more than three dimensions, the Earth would either have fallen into the Sun, there to vaporize in its blazing fires, or would have rocketed off into the near absolute zero of interstellar space. The same fate would have befallen every other planet in the universe. Life could never have developed under these conditions. Had space fewer than three dimensions, the orderly communication essential to the operation of the nervous system and the flow of blood would have been hopelessly scrambled. Evolution, no matter how ingenious, would have faced insuperable difficulties in trying to overcome this obstacle. If the strong interaction were slightly less strong, the Sun, the light of the world, would never have been lit; if the neutron did not outweigh the proton by a tenth of a percent, the Sun and every other star in the sky would quickly have burned out. Evolution would have been helpless to combat such a total absence of stable, long-term sources of energy.

In his *Fitness of the Environment*, Henderson put his finger on the crux of the matter. The passage has already been quoted but it is worth doing so again—with different emphasis this time. "Darwinian fitness is compounded of *a mutual relationship between the organism and the environment*," he wrote, "and the actual environment is the fittest possible abode of life." Certain things evolution can do. It can design a biological clock that ticks synchronously with the passage of the seasons. It can make a fish that swims and a bird that flies. But other things evolution cannot do. It would never be able to design an organism capable of withstanding the cataclysmic explosions resulting from an imbalance between electron and proton charge. It could not operate in that bland and unreactive cosmos in which three nuclei failed to resonate in the red giant stars. The incessant, planet-wide catastrophes that would ensue if the stars were closer together would freeze, vaporize, and batter it into submission.

As emphasized in the Prologue, among scientists today there is a strong dissatisfaction with the notion of a cosmos suited to life—a powerful, intuitive distaste for the idea, a distaste against which I myself have had to struggle often, and which goes far beyond logic. What is the reason for this widespread revulsion? I believe the revulsion is ultimately historical in nature: It is an expression of a residue embedded within our psyches left over from centuries of bitter history. I believe that the minute the subject is raised, the weary feeling descends that all this foolishness has been seen before— seen and battled against, and not just once, but over and over again.

Whewell lived to read Charles Darwin's *Origin of Species*. His reaction to that extraordinary work is not recorded, but that of others of his age is. The theory of evolution burst like a bombshell over the nineteenth century, and it triggered one of the most furious battles in the history of science.

Evolution was widely perceived to be an attack on the validity of the Christian faith. It denied the biblical account of creation as set down in the book of Genesis, and it denied that mankind was a special object of that creation. Darwin was denounced from every pulpit. Perhaps the most famous episode occurred in 1860 at Oxford University, in a meeting of the British Association for the Advancement of Science. There Bishop Samuel Wilberforce, leader of the opposition, debated "Darwin's bulldog," T. H. Huxley. In a heated moment Wilberforce inquired of Huxley whether he was descended from an ape on his grandfather's or his grandmother's side. "The Lord hath delivered him into mine hands," Huxley is said to have whispered to a companion, and he rose to annihilate the opposition:

> If . . . the question is put to me,
> would I rather have a miserable ape for
> a grandfather or a man highly endowed by
> nature and possessed of great means of
> influence, and yet who employs these
> faculties and that influence for the
> mere purpose of introducing ridicule
> into a grave scientific discussion—I
> unhesitatingly affirm my preference for
> the ape.

Some battles are fought and won. Others must be fought and won—and then fought again and yet again. A Tennessee law passed in 1925 declared illegal the teaching of any doctrine denying the divine creation of man. Everyone knows the famous story of the Scopes trial and the courtroom battle between William Jennings Bryan and Clarence Darrow. Less well known is the outcome of that trial: Darrow lost the battle and Scopes was found guilty. The Tennessee Supreme Court blandly noted that any school that found the law hampered its teaching of biology was perfectly free to omit the subject entirely from its curriculum. Scopes was ultimately acquitted on appeal, but the law remained on the books until 1967.

In 1981 a complaint was filed in Sacramento Superior Court charging that the teaching of evolution in California violated children's rights to their own religious beliefs. The court eventually decided against that complaint, but not until untold time and money had been expended in the defense. In the same year the state of Arkansas passed a law requiring the balanced treatment of "creation science" and evolution in public schools. That law too was overturned—but no matter: By the end of that year fully twenty other bills were pending in various states.

It is the biologists who have had to bear the brunt of the seemingly endless battle against creationism. Physicists and astronomers, on the other hand, are more distant from the struggle. But we too have our historical memories. We remember Copernicus, who feared to publish during his lifetime his great and revolutionary theory that the Earth revolved about the Sun. We remember Galileo, silenced by the terror of the Inquisition and forced into house arrest during his last eight years. And we remember Giordano Bruno, who defended the Copernican theory and taught the plurality of worlds, and who was burned at the stake for his teachings.

The scientific revolution constituted an explicit denial of an ancient and universally held idea: the idea that humanity occupies a privileged position at the center of the universe, that in the overall scheme of things we are special objects of attention. This idea is known as anthropocentrism. In the Uffizi Gallery of Florence, one of the great art museums of the world, is to be found an explicit depiction of the anthropocentric view of the universe against which the scientific revolution was forced to struggle. It is a 1413 altarpiece by Lorenzo Monaco depicting the Coronation of the Virgin. Monaco elected to depict this coronation in the center of his canvas. The lower portion, though, he devoted to something else: to a multicolored arch, studded throughout with stars, upon which the figures stand. Beneath the arch are vignettes of earthly life.

That painting is a lesson in Renaissance cosmology. To the Renaissance mind the geocentric cosmology was not just an abstract idea. It held deep religious meaning, and that meaning was explicitly anthropocentric. The Earth, located at the center of the cosmos, was the abode of humanity; surrounding the Earth, and at progressively greater distances from it, were the spheres of the Moon, Sun, and planets; and surrounding these was the sphere of the fixed stars. As one moved outward through all these spheres the general level of sin and imperfection found upon the Earth progressively gave way to the holiness and perfection of the heavens; and finally, out beyond the sphere of the stars, God Himself was thought to dwell. And more than that: As depicted in Monaco's painting, the hosts of heaven were aware of us. They were looking—they *cared*. The scientific revolution was far more than the creation of a new theory concerning the motion of the Earth. It was a direct, frontal attack on a number of doctrines of the Christian faith.

Everyone knows the story of Galileo, and his prosecution by the Church for his defense of the Copernican system. Less well known, but far more terrible, is the story of Giordano Bruno. Bruno too was an ardent champion of the Copernican theory. He went beyond Copernicus, however, in also affirming the infinity of the universe and the plurality of worlds. Bruno was born in 1548 near Naples, the son of a professional soldier, and he seems to have been a brilliant, restless, endlessly disputatious man. He took religious instruction but read a forbidden text denying the divinity of Christ and was accused of heresy, fled to Rome but found himself (unjustly) accused of murder, was excommunicated and fled again. Two years later he was in Geneva employed as a proofreader. There he embraced the Calvinist faith; after publishing an attack on a Calvinist professor he was arrested, excommunicated, rehabilitated, and allowed to leave. He moved to Toulouse and then to Paris, where he published a play attacking the corruption of Neopolitan society, then to London and finally Oxford where, in a series of lectures in the summer of 1583, he defended the Copernican doctrine. Finding the reception hostile he moved again to London, where he wrote *The Ash Wednesday Supper*, a series of dialogues in which he maintained his belief in the motion of the Earth, the infinity of the universe, the plurality of worlds, the pedantry of his detractors, and the overall inadequacy of English society. Over the next seven years Bruno crisscrossed Germany and France, propounding the peaceful coexistence of all religions based upon the open discussion of their differences and picking up yet another excommunication, this time from the Lutherans, along the way.

Bruno's fatal mistake was his return to Italy in 1591. There he was denounced to the Venetian Inquisition for his theories. The trial appeared to be going well at first, but then the Roman Inquisition demanded his extradition. In 1593 he entered the Roman jail. The trial lasted seven years, and on February 8, 1600, the Inquisition condemned him to be burned at the stake. The sentence was carried out nine days later in the Campo de' Fiori in Rome.

Did Bruno sleep the night before his burning? Did he weep? Was he weeping as they dragged him from the cell and into the square where he saw for the first time the stake, weeping as they bound him to it and lit the flames? As the fire reached him his tongue was in a gag, but Bruno's cries were heard. They were heard not just throughout the square that bleak winter day, not just throughout Rome, but throughout the Western world. And I believe I can hear them even now, a faint tenuous keening amid the rustling of bushes in the wind.

The concept of a universe fitted for life has arisen at the end of a long and complex historical journey. We are all inheritors of the burden of that journey, and this, I believe, accounts for the well-nigh universal rejection of that concept by scientists today. But I would plead for a fairer hearing. This new view's resemblance to the outmoded exhortations of natural theology is apparent only, a chimera, and should not be allowed to obscure the radical differences between the two. Its resemblance to the ancient doctrines against which the authors of the scientific revolution struggled is likewise merely apparent. I am not blind to the political context in which modern science takes place. I am not blind to the widespread contemporary belief in astrology, and to the incessant efforts of creationists to inject their religious beliefs into the teaching of science. I understand full well the yet greater political problems we would face in our combat against creationism were my thesis taken seriously by the scientific community. All this I grant. But it is still no reason to turn back from the potential fruits of discovery.

10

The Moment of Creation

Sigmund Freud reminded us that the child is father to the man. By this he meant that within us all we carry traces of our earlier existence, and that these traces are often of overriding significance. The same is true of the universe as a whole. Its present structure evolved from a prior one, and this in turn from yet an earlier structure, and so on back to the very moment of creation itself—to the Big Bang in which the universe appears to have begun. As the archaeologist finds relics of the past buried beneath the sands, so the astronomer finds traces of creation.

Cosmology is the science of the universe as a whole. It is a subtle subject, and a complicated one. But some things about it are not subtle. Rather than being complex and obscured in tangles of detail, they hit you in the face. Remarkably enough, though, for decades few scientists paid any attention to these things. It was a little like trying to predict the weather but failing to ask why winter is colder than summer.

Perhaps it was a matter of missing the forest for the trees, of needing to back away from the complexities of cosmology and survey the subject afresh.

When one does so, a whole set of enigmas arises. These enigmas refer to the moment of creation, to the Big Bang. There are four, and they are deeply mysterious. Furthermore, it turns out that each one is intimately related to the possibility of our existence. The Big Bang, that explosive instant, was delicately and precisely tuned—tuned in four different ways. Had it not been for that tuning, life could never have arisen in the universe.

Cosmology has existed as long as humanity, though initially it was purely speculative in nature. The scientific revolution provided us with the means of beginning a realistic study of the subject. But cosmology remained in its infancy until 1929, the year in which the American astronomer Edwin Hubble made one of the most remarkable discoveries of the century: that the universe is expanding. Steadily, ponderously, every galaxy is drifting away from every other. That expansion is now known to be the remnant of the Big Bang, the explosive, all-encompassing cataclysm in which the universe began.

What is the ultimate fate of the expansion? One possibility is that in the long run it will reverse. Untold ages hence, in an evolution vast, ponderous, and majestic, the galaxies' outward motion will slow to a halt. Briefly—"briefly" in cosmic terms—they will hover. And then, slowly at first but with ever-increasing speed, they will plummet together.

When the overall scale of the universe has contracted by a factor of fifty, these galaxies will merge. Their complex and lovely structures will be destroyed as they crush into one another. The nighttime sky in those far-distant epochs will be spectacularly beautiful as a multitude of stars, some belonging to our Milky Way Galaxy, some to others, mingle together. New constellations will appear in the heavens. At the same time, however, winters will begin to grow uncommonly mild, summers uncommonly severe. A new form of heat will be flooding down—heat not from the Sun, but from the universe itself.

The Big Bang was blazing hot. An intense light filled all space, the shine of that terrible fire. At this very moment that light still exists, a relic of our birth, but diluted in its intensity by the subsequent expansion, and its wavelength stretched by the expansion so that what was once visible light now is a faint emission detectable only by radio telescopes—the so-called cosmic background radiation. But if the universe ultimately begins contracting this process will reverse: The cosmic background's intensity will grow and its wavelength contract back toward the visible region of the spectrum. Also it will heat the Earth, and every other object in the universe.

As the universe continues contracting it will grow sufficiently hot that oceans will boil. The background radiation will gradually shift to a sullen, deep-red glow, ominously filling the nighttime sky as it pours down upon us. As the contraction proceeds the radiation will grow brighter and shift in color up the spectrum, ultimately reaching a hideous electric blue. Under its awful heat rocks will melt, the surface of the Earth grow molten—and then grow hotter still, the very planet itself vaporizing, dissolving into the fire. The same will be the fate of every other body in the universe: first the moons, asteroids, and planets, eventually the stars themselves. In their final epoch these stars will be whizzing by each other at ever-increasing velocities, occasionally colliding, but more often undergoing near misses as they crowd together. Ultimately they will be destroyed not by collisions, but by the heat of the cosmic background.

Still more compaction. The heat grows, the density. Things happen at an ever-increasing rate as the universe collapses in on itself. A century before the final catastrophe the temperature is 100,000 degrees; it reaches a million degrees one year before the end, and a billion degrees one minute before it. In the final instant every object in the universe—the shredded remainder of every star, of every planet, every atom and speck of dust in the farthest reaches of space—is consumed, crushed into every other object, crushed into itself and annihilated in the ultimate cataclysm. The storm of that destruction will be a mirror image of the fires of our birth, the Big Bang in which the universe began so many eons ago.

But this is not the only possible outcome of the present expansion of the cosmos. There is a second possibility: that the expansion will never reverse, but will continue forever.

All eternity stretches before us in this case, a limitless expanse of time. The galaxies will continue flying apart from one another, forever slowing but never halting, in a vast outward motion without end. Billions of years from now the Sun will go out, its nuclear fuel consumed. Once this happens it will gradually contract, ultimately stabilizing as a so-called white dwarf star. The output of energy from such a star is very low, so the Earth, feebly warmed by its faint heat, will grow bitterly cold. Above the frozen landscape, above the endless drifts of snow—snow on the Sahara, snow on what had been the Caribbean Sea—tiny, pointlike, blue-white, rising in the morning but so dim that stars can be seen in daylight, above it all will shine the meager remnant of what once had been the Sun.

Other suns will shine in that sky, some ultimately ending their histories as white dwarfs, others more violently as pulsars, yet others collapsing to form black holes. But no matter what its particular fate, each sun will eventually go out. Ages hence the nighttime sky will be almost entirely empty. The Milky Way, the constellations, will have vanished. In the near-perfect blackness of the night two, five, perhaps ten stars will faintly shine; and as the ages pass they too will grow yet dimmer, the sky yet more vacant, the cold yet more severe. Only on some rare planet huddling close to its sun might some remnant of civilization conceivably remain.

If so, these survivors' conception of the cosmos will differ utterly from our own. At the present epoch in the expansion of the universe, certain galaxies are close enough to be seen with the naked eye. The Andromeda Nebula and the Whirlpool Nebula are examples, and residents of the Southern Hemisphere have an even better view of the two closest galaxies, the Clouds of Magellan. A backyard telescope reveals more, and with the largest, literally billions of distant galaxies can be found.

But as the universe expands all grow faint and are carried

off into space. Ages hence they will have dimmed to invisibility and drifted beyond the reach of even the greatest telescopes, and any scientists of those times will have no evidence of their existence. They will tell of a universe empty save for one sparsely populated Milky Way Galaxy, an island in the infinite ocean of the night. Of the cosmic expansion, of the Big Bang and our explosive origin, they will have no clue. Ultimately the last star will burn out. And then, lifeless, dark, steadily cooling toward a temperature of absolute zero and filled with burned-out cinders, the galaxy will continue rotating ponderously upon its axis, steadily slowing in this rotation, the other galaxies as well, all invisible; and every one of them drifting endlessly, limitlessly, outward into the abyss.

Those are the two possibilities.

How to decide between them? Which will be the actual fate of the universe: eventual recollapse or indefinite expansion? An analogy to the cosmic expansion is found in the simple act of throwing a stone upward. Throw it slowly and the stone falls back to Earth. Throw it rapidly—by firing it off in a rocket, for instance—and the stone never returns, but continues on its journey without end.

It is gravitation that determines the final outcome. The question is whether gravity is powerful enough to reverse the stone's flight. In cosmology the concern is with the gravitational tug of the universe as a whole. This is determined by how tightly the matter within the cosmos is packed—by the overall density of the universe. A high density implies a strong gravitational tug, sufficiently strong to reverse the present expansion at some point in the future, convert it into a contraction, and ultimately lead to annihilation in a second Big Bang. Alternatively, low density implies a weak gravitational tug, insufficient to reverse the present expansion, which in this case will continue without end. The dividing line between these two possibilities is known as the critical density.

As of this writing, we have not yet succeeded in determining the actual density of matter in space with sufficient accuracy to learn whether it exceeds this dividing line or not. We do know that it is certainly close. The question therefore remains open. But as emphasized above, the important thing is to back away from the complexities of cosmology and survey the subject afresh. When one does so, an extraordinary insight emerges. The future turns out not to be the point. *The question is not whether the universe will eventually recontract into a second Big Bang. The question is why the universe did not do so the instant after it was created.*

Don't ask about the future. Ask about the past. The present stage in the expansion arose from a previous one. Begin by fixing attention on some particular benchmark—any benchmark: 5 billion years ago, say. Clearly, at that far-distant epoch the cosmos was expanding in such a manner as to ultimately reach its present configuration. What was that manner?

It must have been expanding more rapidly back then, for like the stone flying upward, the cosmic expansion steadily slows. Also the galaxies must have been closer together—the universe must have been more dense. Of course, it could not have been *too* dense, for the more compact the cosmos, the stronger the gravity opposing its expansion, and the sooner it was destined to recontract. There is at least one thing we know about the universe in those times: Its density must have been such as to allow it to survive till now . . . and 5 billion years is a long time.

It turns out that the density of the universe could have been no greater than eight times the critical density back then. Any greater and it would have quickly annihilated itself in a second Big Bang. But surely this is not so difficult to imagine. What is the problem?

The problem is that five billion years ago does not mark the beginning of the universe. It is simply a convenient benchmark from which to reckon things. That epoch arose from a still earlier one, in which the universe was expanding yet more rapidly, and was yet more compact. How compact?

Which epoch do you want to discuss? Try a point in time 1 million years after the Big Bang. In that unimaginably distant age the cosmos was ferociously hot, rushing outward at an enormous rate, and it was exceedingly compressed. But once again, it could not have been too compressed. The universe must have been no more than one third of 1 percent greater than critical density at that stage. Any greater and it too would rapidly have been annihilated in a bang.

That's a fairly precise adjustment. But of course there is nothing sacred about the epoch 1 million years after the Big Bang. It, in turn, arose from a still earlier configuration. Push back still more. Push back to one year after creation, to the very infancy of the universe. It was no more than 0.00003 percent denser than critical. But why stop there?—keep going. When the universe was one hour old it could have exceeded critical density by no more than 0.00000008 percent; the slightest deviation would have caused it to recollapse essentially at once. And the further one goes, the more perfect does the adjustment become.

In the limit, we come face to face with the inconceivable, the unfathomable: the moment of creation itself. Suddenly we know something about it. It was such as to allow the present configuration of the cosmos ultimately to arise, and this required an adjustment not of one part in a thousand, not of one part in a trillion, but of one part in infinity. Creation was perfect.

Had it not been so, the cosmos would have winked out of existence the instant after it had been created. But for life to develop time is required, and lots of it—time for the long, winding course of cosmic history to blossom forth in the development of complex molecules, time for these molecules to congregate together and form the first primitive organism, and time for evolution to produce higher forms of life. We are the products of 17 billion years of cosmic evolution. But failing that mysterious perfection of the moment of creation, those immense gulfs of time would not have been available; neither we nor any other form of life could ever have come into being.

This analysis, of course, is predicated on an assumption—the assumption that the universe in fact is destined to recollapse. The opposite also remains a possibility, and if at the moment of creation the density was less than critical, no problem with destruction in a second bang would arise. On the other hand, recall that observations place the present density at least close to critical; in order to arrive at this crude matching today, a perfect matching was required at the creation. It seems that there was a squeeze, a literally perfect adjustment from both sides.

That adjustment is the first enigma. For reasons that will become evident in the next chapter it has a name: *the flatness problem*.

The second enigma relates to the overall distribution of matter in the universe. How is the cosmos organized?

At the level of our experience the fundamental unit of organization is the Earth. With the rare exception of meteors, every object we encounter belongs to this planet. Of course there is also the Sun, not to mention meteors and the other planets, moons, comets, and so forth. Together all these form a single structure, the solar system.

There are other stars in the sky, each possibly with its own planetary system. Lift off now in some hypothetical rocket, an interstellar explorer, and voyage thousands of light-years into space. Look back and survey things in the general vicinity of the Sun. How are the stars organized about us? Figure 28 gives a good representation of the view backward from that hypothetical spacecraft. In that view the Earth, the Sun, the Moon—all have receded into invisibility. But a pattern to the organization of stars has become evident. They form a tube, a gigantic arcing spiral arm.

Travel yet farther. Survey space over scales of hundreds of thousands of light-years. Now the arm is revealed to be part of our Milky Way Galaxy, a disk shot through with winding arms similar to that shown in Figure 29. Other galaxies float in the void, some spiral like our own, others spherical or elliptical in shape. Launch the cosmic voyager into the vastness of intergalactic space, and after a journey of tens of millions of light-years look back. Our galaxy and its neighbors are seen to be organized into a loose cluster, the so-called Local Group. Yet other clusters of galaxies populate

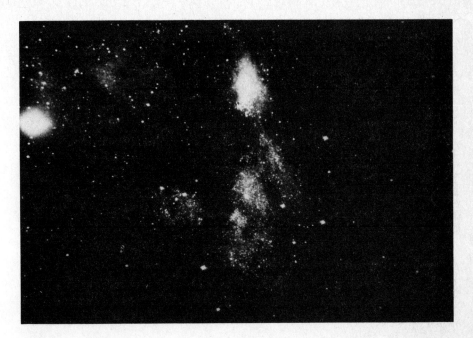

Figure 28: A photograph of the outskirts of the Andromeda Galaxy showing its spiral structure.   *Mount Wilson and Las Campanas Observatories, Carnegie Institution of Washington*

the sky. And finally, the distribution of matter in the cosmos over the vastest scale we are capable of surveying is shown in Figure 30. Here the clusters themselves are grouped into yet larger units termed superclusters. Additionally, a glance at that figure shows there to be a faint hint of some stringlike or bubblelike pattern over these inconceivable distances. Galaxies in large numbers seem to arrange themselves into filaments, the filaments quite poorly defined but apparently real units of organization, twisting and wandering through space. Also on this scale the universe appears to resolve itself into a spongelike pattern, the filaments and superclusters alternating with relatively empty bubblelike voids. And as we construct yet greater and greater telescopes, and as we probe yet deeper into space, no limit to this incomparable swarm has ever been found. So far as we are able to judge, no planet exists from which an edge of the cosmos might be glimpsed.

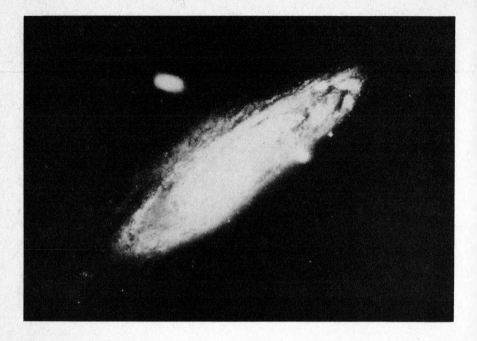

Figure 29: The Andromeda Galaxy. (Note the satellite galaxies lying along-side.)   *Palomar Observatory*

That is the organization of the universe. An important feature is that it is pretty much the same everywhere. Clusters, superclusters, and filaments lying in one region of space are no different in kind from those lying somewhere else. The overall distribution of matter in our general vicinity is identical to that far away. *When surveyed over great distances, the universe is uniform.*

A grand picture, extraordinary in its scale . . . but what's the problem? The problem is that, as before, this structure must have evolved from a previous one. Project backward in time, and ask what the distribution of matter must have been like in the past.

It must have been *smoother* at every point in the past, for the simple reason that gravitation, the dominating influence in the universe, magnifies whatever irregularities may exist. Any concentration of matter tends to attract toward it yet

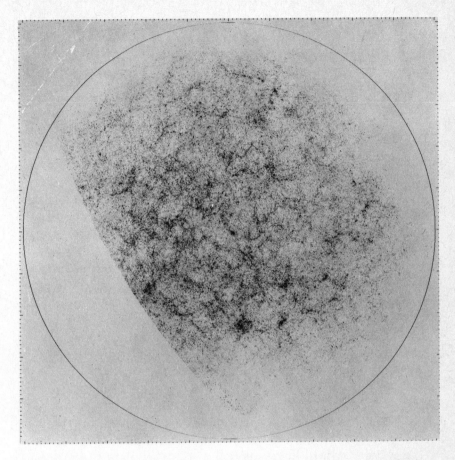

Figure 30: One million galaxies: a computer-processed illustration in which the location on the sky of each galaxy is indicated by a dot. The line across the lower left is the horizon limit of the Lick Observatory, where the data were taken. More recent research, in which the distance of each galaxy is also studied, has confirmed the spongelike nature of the distribution that is suggested here.   The Astronomical Journal

more matter, thus increasing the degree of concentration. In this regard the cosmos is akin to some immense pudding, slowly coagulating as it cools. If the pudding contains lumps they will grow as time passes; similarly, had the universe contained any tight knots in the past, strong irregularities in its overall structure, they would have been drawn together by gravity into far stronger irregularities today—and these we do not find. The implication is they were not present back then.

Projecting back still further, the same argument shows that at very early times the distribution of matter in the universe must have been quite remarkably smooth. How smooth? Again, the answer depends on which stage one is thinking about. But the flatness problem has already shown the overall density of matter to just equal the critical density of the universe. Imagine then the slightest, the most infinitesimal of knots shortly after the Big Bang, corresponding to a tiny increase of this density in some region of space. What would have been its eventual fate? Since the overall density was just critical, the knot must have been *denser* than critical. It would have separated out from the general expansion, reversed and recollapsed . . . not into a second Big Bang, but into a gigantic black hole.

A disturbance of the opposite kind, a region of space less dense than average, would have been below critical density, and so would have continued expanding forever—but very rapidly, and by now would have become an immense void, empty even by cosmological standards. The conclusion is that had such knots and rarefactions been present at the instant of creation, the universe today would have been unrecognizable. It would have consisted solely of giant black holes wandering through otherwise empty space.

Such a cosmos would have been utterly unfit for life, containing no galaxies, no stars with their life-giving light and warmth, and no planets upon which that life might flourish. On the other hand, it is difficult to understand why those knots were absent at the moment of creation. The tiniest of disturbances, bumps, or shudders would have produced them, and the Big Bang was an exceptionally violent event. Nevertheless it appears to have been less like an explosion than like some immense river, its pounding, hurried flow magically held mirror-smooth. And it is from the perfection of this flow that we have our being.

That is the second enigma. It has a name: *the smoothness problem.*

*   *   *

Next is *the horizon problem.*

Survey the universe out to great distances in some direction—any direction. Now survey to the same distance in precisely the opposite direction. As stressed above, the same general pattern will be found.

Observations of the cosmic background, the residual glow of the Big Bang, enable us to conduct this survey out to very great distances indeed. This background radiation was emitted soon after the bang (where "soon" is to be taken in cosmic terms). So when a telescope receives this emission, it is receiving something that has been flying toward us for essentially the entire history of the universe. How long has that been? We do not know very well, but 17 billion years is a good ballpark figure. And clipping along as it does at the speed of light, the emission has traveled 17 billion light-years in that amount of time. Observations of the cosmic background plumb the universe to this great distance, incomparably farther than we can see by any other means.

Repeat now with this background the observation described above: Observe it first in one direction, then in just the opposite. The glow is found to have the same intensity in both cases—not just crudely, not just more or less, but the same intensity to a very high accuracy. The glow, being heat radiation, measures the temperature of the Big Bang, and it demonstrates the bang to have been equally hot in those two directions.

But it could not have been. That is the enigma.

For after all, most things have differing temperatures unless some process has actively intervened to equalize them. Brazil is warmer than Vermont, Mars colder than Venus. In both examples the temperatures differ because the two locations are far from one another. And in cosmology, the two locations are exceedingly far apart indeed.

Figure 31 diagrams the situation. On it, A marks the location of the first region surveyed, B the second. As emphasized above, because their signals have been traveling for 17 billion years at the speed of light, each must be 17 billion light-years from the Earth—34 billion light-years from each other. That, of course, is a very great distance indeed. It is, in fact, too great a distance for any conceivable process to act that might have equalized their temperatures.

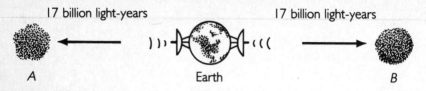

17 billion light-years   17 billion light-years

A   Earth   B

Figure 31

What could that process have been? One possibility is mixing. If two objects are brought together and stirred they reach the same temperature (imagine the first to be a hot cup of coffee, the second a cold spoonful of cream). Similarly, some process might have mixed A and B together, thus equalizing their temperatures at a remote point in the past, and then separated them to their present great distances. A second possibility would be the steady conduction of the greater heat of one region into the lesser heat of the other, so evening out the difference; a third would be the transfer of heat via radiation.

These are the only ways in which the temperatures of regions A and B could have been equalized. But there is a limit, placed by the theory of relativity, on how rapidly any of them could have acted. According to relativity, nothing can go faster than light, and in placing this restriction the theory does not just refer to spaceships and the like. The restriction also applies to the rate at which two substances can be mixed, the rate at which heat can be conducted, and the rate of transfer of heat via radiation.

Relativity's restriction is that it must have required at least 34 billion years to equalize the temperatures of those two regions 34 billion light-years apart. But that huge expanse of time simply is not available. The very universe has existed only half so long. Some mysterious process must have brought the temperatures of A and B together, and that process, whatever it was, must have been capable of moving at twice the velocity of light.

In fact it must have moved faster still. Return to Figure 31. There is an ambiguity in this diagram, an ambiguity having to do with time. When does it refer to? As for the detection of the cosmic background, that happens now. But as for its emission—that happened a long time ago. It happened during the Big Bang. Figure 31 is a diagram of the universe in the far-distant epoch of its creation. It was then that A and B were of equal temperature. It was then that they were 34 billion light years apart. How they could have reached such an enormous separation in so short a time is discussed in the next chapter, but it is not the point here. The point is that the agency that equalized their temperatures did not have 17 billion years in which to operate at all. It had a far shorter stretch of time. It must have traveled not twice, but thousands of times faster than light . . . which is quite impossible.

Marco Polo voyages to a distant land. He journeys far beyond the horizon—not just our familiar, relative horizon, but an absolute one, fixed by the laws of nature. No European has ever been there; nevertheless, upon arriving he finds a place identical in every respect to home.

That is the horizon problem. It is related to the smoothness problem, for if any two regions differed in temperature they also would differ in pressure. The resulting imbalance would force neighboring regions into one another, quickly spoiling the smoothness. Thus our existence depends not just upon the perfect uniformity of the distribution of matter at the moment of creation, but on the uniformity of its temperature as well.

Last is *the matter-antimatter problem.*

In 1932 the American physicist Carl Anderson discovered a new subatomic particle. It had the same mass as the electron but the opposite charge. Nothing so very remarkable there; the proton's charge also was the opposite of the electron's. The surprise came when the new particle encountered an electron. Both vanished utterly, and in their place appeared a brief but intense flash of light—a double flash, in fact: two photons, or particles of light. The reaction is diagramed in Figure 32.

Anderson's particle is now known as the antielectron, and the reaction shown in Figure 32 as annihilation. In such a

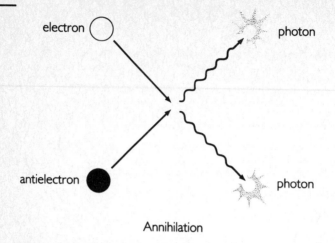

Annihilation

Figure 32

reaction matter is entirely annihilated, and light, which is pure energy and which contains no matter whatsoever, is created. The reaction represents the complete conversion of matter into energy, as expressed in Einstein's famous $E = MC^2$. To every particle there corresponds an antiparticle capable of annihilating it: To the proton corresponds the antiproton, to the neutron the antineutron. Each antiparticle is opposite in charge to its corresponding particle, but identical to it in every other way.

That identity is crucial. Subatomic particles in aggregate form matter. Similarly, antiparticles in aggregate form antimatter, and the stuff so formed behaves identically to the corresponding matter. If water is H_2O, antiwater would be anti-H_2-anti-O; it would look like water and flow like water, would boil at the same temperature and freeze to make antisnow.

The identity strongly suggests that when the universe was created, it was created with equal quantities of matter and antimatter. Since the underlying physics of the two is identical, their production in the Big Bang ought to have been identical too. But the present universe is not constructed in this way at all. Rather, it is made solely of matter. In ordinary circumstances antimatter is never found; indeed, the stuff appears only briefly, in the debris of collisions between high-energy particles, and once produced it quickly annihilates. The fundamental identity between the two is violated in the present structure of the universe.

On the other hand, we owe our very existence to this violation. For suppose equal quantities had been produced in the Big Bang. Figure 33 would then be an accurate diagram of the content of the primordial cosmos. Had this been so, everything would have quickly annihilated. Paired as it was with a corresponding antiparticle, each particle would have succeeded in destroying itself. After a brief interval of universal annihilation, no matter of any sort would have survived. Nor would any antimatter. The ultimate configuration would be that shown in Figure 34. It consists of nothing but radiance: pure light, with no admixture of matter at all. And as for that light, with no material objects around to absorb it, it would survive forever.

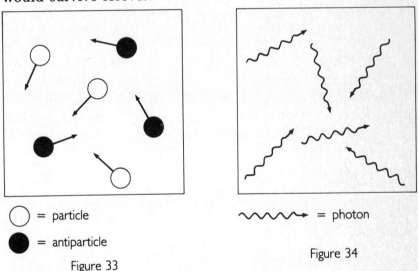

◯ = particle

● = antiparticle

Figure 33

〜〜〜➤ = photon

Figure 34

In such an eventuality, the present-day cosmos would not be composed of *things* at all. There would be no stones and trees, no planets, meteors, or suns. Instead of all the rich diversity of material objects that actually compose reality, there would remain merely a uniform, utterly steady shine of light: perfect radiance.

A striking prospect, poetic in its beauty—but not conducive to our existence, nor that of anything else. Clearly the early universe must have been imbalanced in its content, containing more particles than antiparticles, as illustrated in Figure 35.

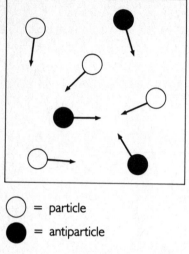

= particle

= antiparticle

Figure 35

In this instance each of the antiparticles would have been consumed in the annihilation reactions, but the few unpaired particles would have survived unscathed. After the termination of the annihilation stage, the configuration would have been as in Figure 36.

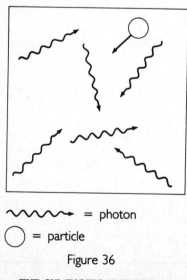

= photon

= particle

Figure 36

The photons would have remained, ultimately to become the faint glow of cosmic background radiation we observe today. But something else also would have remained—those few remaining bits of matter. Out of them the entire material universe was destined to be constructed: the stones and trees, the planets, meteors, and suns . . . and us. We are the descendants of a primordial imbalance.

How great was that imbalance?

Figure 32 diagramed the annihilation reaction in which particle and antiparticle vanish to produce two photons, or particles of light. The reverse reaction is also possible, as represented in Figure 37. Here two photons collide to produce a particle and an antiparticle. Matter is created from energy; the process is termed pair creation.

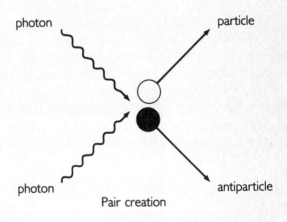

Figure 37

Just as the light emitted by annihilation reactions is very intense, so pair creation will occur only if the colliding photons are highly energetic. Shining two flashlights upon one another will not do the trick. The light must be brilliant, and its wavelength must be short—gamma-ray wavelengths are required. In ordinary circumstances, therefore, pair creation does not occur. But the Big Bang was not ordinary. It was flooded with light, the shine of creation, and that extraordinary radiance was capable of initiating pair-creation reactions. In this way the early universe was flooded with particle-antiparticle pairs. It remained hot enough to maintain these pairs for the first second of its existence; and after that point there occurred the universal annihilation described above, which led to the present-day content of the cosmos.

By everything we know, this early stage in the evolution of the universe, produced by pair-creation reactions, should have maintained the correspondence between matter and antimatter illustrated in Figure 33 above. Apparently, however, it did not. For some mysterious reason the configuration was rather as illustrated in Figure 35. But this figure is schematic only, and it fails to give an accurate impression of the magnitude of the lack of correspondence.

To gain that impression, imagine that a census is taken of Chicago and that this census reveals the city to contain just as many women as men. A striking result; intrigued, the census takers decide to expand their sphere of operation, and survey the entire state of Illinois. Remarkably enough, here too they find as many of one sex as the other: every woman paired with a man, every man with a woman. Washington is informed, and eventually the census is extended across the country. And in every city of the nation, in every county and state, the same pattern is found . . . until one day, sifting through the data for a small town in central Nevada, an imbalance is discovered—one unpaired woman.

This single anomaly out of the entire population of the United States is a good representation of the degree of imbalance of the cosmos in the fires of its creation. That lone unpaired woman is mother to us all. It is difficult to know which is more surprising: the fact that the imbalance existed, or the fact that it was so very small. Both are mysterious. It is hard to avoid the impression that the moment of creation "tried" to achieve a state of perfect correspondence in which to each particle of matter there corresponded one of antimatter—and that it failed. If so, we owe our existence to that failure.

II

Grand Unification and the Inflationary Universe

But after all, who knows, and who can say
whence it all came, and how creation happened?
The gods themselves are later than creation,
so who knows truly whence it has arisen?

Whence all creation had its origin,
he, whether he fashioned it or whether he did not,
he, who surveys it all from highest heaven,
he knows—or maybe he does not.

—Rig Veda (c. 1500 B.C.)

Creation is the great unknown. As the words of the Rig Veda teach us, before its solemn mystery little is left other than to refrain from uttering nonsense. For all its strides, modern science has made not the slightest progress in comprehending the very act of creation itself.

For years, then, the great cosmic enigmas, upon which the existence of life in the universe depends, appeared beyond the reach of analysis. What hope could there be of coming to grips with the flatness, smoothness, horizon, and matter-anti-matter problems, bound up as they were in that great mystery? But the ongoing drive of research is restless, incessant, and the conviction among scientists is universal that all things obey fixed and immutable—and comprehensible—laws. Somehow, everyone always knew that ultimately an answer would be found.

This answer may now be in hand, and in the present chapter I wish to describe it. Perhaps most remarkable of all is how our new understanding has been won. It has not been won as the result of some increase in our understanding of cosmology. It has not been won as the result of some profound advance in our understanding of creation. Rather it appears that the four cosmic enigmas involve processes occurring not at, but very shortly after the moment of creation, and that these processes involve a field of knowledge apparently unrelated to cosmology: elementary particle physics. The nature of matter at the level of its very smallest units turns out to have enormous effects on the nature of the cosmos as a whole.

Nowhere is the life-giving quality of the laws of physics more evident than in the case of the universe as a whole. What relation could there be between life on the one hand and the ultimate building blocks of matter on the other? Certainly cosmologists do not concern themselves with such questions. Neither do elementary particle physicists. Nevertheless, as a result of their joint efforts, we now have before us a case study, an explicit example in which can be seen the remarkable fashion in which wildly disparate laws of nature conspire to bring forth life.

In some ways it is hard to see why the matter-antimatter problem should be thought of as a problem at all. After all, the great flood of particles produced by pair-creation reactions in the early universe yielded a state of near-perfect symmetry. The disparity was tiny—a mere one part in hundreds of millions.

And indeed, why count so carefully? Why not allow a little fuzziness in the census? In that case the problem would never have arisen at all. Furthermore, each one of us lives with such imprecision all the time. Not long ago a brilliant red sock appeared in my laundry. I had never seen the damned thing before in my life, but the fact did not bother me. I simply tossed it into the drawer and, sure enough, noticed several weeks later that it had vanished back into the limbo from which it came. And as for the much-vaunted precision of financial accounting, it is a myth: Not once have I been able to balance my checkbook successfully. Dollars and cents come and go like so many will-o'-the-wisps.

But nature does not allow such looseness. Nature keeps track, and it does so with absolute, unbending precision.

What nature keeps track of is baryon number. Baryons—the term comes from the Greek for "heavy"—are heavy subatomic particles, and baryon number is simply a device for counting them. Thus the proton and neutron are baryons whose baryon number is one, while the electron, a far lighter particle, is not a baryon and carries zero baryon number. An assemblage of two protons and two neutrons, the helium nucleus, has a baryon number of four. Antiparticles, on the other hand, have opposite baryon numbers from their corresponding particles: An assemblage of two protons and two antiprotons contains four objects, but its baryon number is two added to minus two, or zero.

As for light, it is energy, not mass, and its associated particle, the photon, is not a baryon. The pair-creation reaction, then, begins with a state of zero baryon number—pure light—and ends with a state of zero baryon number—a plus added to a minus. It is worth noting that baryon number is not changed by the reaction. Furthermore, the reader can easily verify that the same turns out to be true of every other reaction diagramed in this book. It is true of every reaction that has ever been observed in nature. The urge is irrepressible to summarize all this experience in a law: *Nothing ever happens to change the net baryon number of the universe.*

This principle is known as the law of conservation of baryon number. It is nature's way of counting. And it is why the matter-antimatter problem is a problem. For according to this law, whatever the net baryon number of the cosmos at the instant of creation, nothing that happened since could have altered it in the slightest. It has been fixed, immutable, an absolute property of reality.

The symmetry of physics with respect to matter versus antimatter strongly implies that, when the universe was created, it was created with equal quantities of each. The net baryon number, then, was zero. On the other hand, had that been all there was to it, we never would have come into being, for the subsequent annihilation stage would have left the universe devoid of all substance. Apparently some process must have intervened to bring something out of nothing—to create baryon number, to bring into being the mysterious prevalence of matter over antimatter in the cosmos. This process, whatever it was, must have occurred long before the annihilation stage, in the very earliest phases of the history of the universe. It was this that produced the asymmetric state diagramed in Figure 35 of the previous chapter. The subsequent annihilation, then, would have proceeded in the normal fashion, leaving the baryon number unchanged. Very well, then—it must have happened. The only problem is to discover what it was. Let us call the process baryosynthesis.

But baryosynthesis could not have happened. The law of conservation of baryon number forbids it.

Furthermore, there are powerful reasons for believing in this law. Quite aside from the wealth of experimental data in its favor, the very existence of the world speaks to its truth. If baryon number could have increased shortly after the Big Bang, there is nothing to prevent it from decreasing thereafter—from decreasing right now. In such a case we would *dissolve*. Our very bodies would evaporate into nothingness, for we are composed of matter, and matter is composed of baryons. The stability of matter, the tough, unyielding permanence of substance, depends upon the law of conservation of baryon number. The very thing that might make baryosynthesis possible, and so allow our existence, would undo our being from the other direction.

We are caught in the jaws of a vise. But perhaps there is a

middle ground. Perhaps baryon number is conserved not absolutely, but just most of the time. In such a case, the way lies open to escaping the vise.

Two thousand feet underground, in a salt mine near Cleveland, people are searching for that middle ground.

They are searching for proton decay. Many subatomic particles decay without violating the law of conservation of baryon number—the neutron, for instance, which as discussed in Chapter 7 spontaneously transforms itself into a proton. That law, on the other hand, forbids the decay of the proton, for it is the lightest of all baryons. There is nothing for it to transform into. But if baryon number were capable of changing, all bets would be off. The proton could dissolve into lighter, nonbaryonic particles.

The proton decay experiment is a collaboration between the University of California at Irvine, the University of Michigan, and Brookhaven National Laboratory: the IMB group. Even before they began, the group knew they had a hard job ahead of them. They knew that if the proton decayed at all, it did so very slowly—for there they sat, 17 billion years after the creation of the universe, with all their protons intact. The rate of transformation was therefore less than once per 17 billion years. Indeed it was rarer than that, for Maurice Goldhaber, a member of the group, had realized that if the process occurred it would constitute a form of natural radioactivity. The products of the dissolution of the protons within one's very body would rip through living tissue, causing cancer, genetic disorders, and the like. "We know in our bones it does not decay very quickly" was how Goldhaber put it. In such a way the IMB people knew that the proton, if it decayed at all, did so less than once per 10,000,000,000,000,000 years.

It was clear that anybody wanting to search for proton decay was going to have to do a serious job of it. The IMB group sat down at the drawing board and came up with a set of plans for one of the most ambitious experiments ever contemplated. Whatever the proton dissolved into would be left traveling very rapidly. Such particles speeding through water would emit faint flashes of light, much as a rapidly moving airplane emits a sonic boom. The way to spot the process would then be simple enough: Fill a tank with water and look for those flashes. As for the decaying protons, they could be provided by the water itself: Nuclei of both the hydrogen and oxygen atoms in an H_2O molecule contain them. To have a chance of observing such rare decays, enormous numbers of these molecules would be required; the tank would have to be very large. Furthermore, each flash would be exceedingly faint. Each burst of light would be no brighter than that from a flashlight perched on the Moon. So the water would have to be free of all obscuring sediment. It would have to be of marvelous transparency. Very well, then—filter it. They boned up on the most perfect, the most exquisite techniques of filtration known, and then went on to invent a few more of their own. To detect the flashes they chose phototubes, the phototube being a device of marvelous sensitivity, capable of throwing a loop about the ankles of the most microscopic of flickers, the most tenuous of sparkles, and hauling it in out of the darkness. Each phototube was a triumph of technology. Also, each was expensive. Plans called for 2,048 of them. Finally, there was sure to be a good deal of noise—bursts of light from other sources—jamming the signal they were attempting to detect. Noctilucent organisms were a potential problem, but these would be removed by the very filters that purified the water. A far more serious difficulty would be light emitted by rapidly moving particles from sources other than decaying protons. One such source would be cosmic radiation, the incessant hail of high-energy particles pouring down from the sky. These came from a variety of astronomical sources—pulsars, quasars—and while not medically dangerous would flood the experiment with unwanted signals. The straightforward solution would be to erect shielding about it. On the other hand, so sensitive were the phototubes, so microscopic the signal they were straining to detect, that enormous quantities of shielding would be required. Better would be to let the very body of the Earth constitute the shield—to bury the experiment. They decided to bury it two thousand feet underground. The Morton Salt Company was interested in testing out a new digging device and agreed to excavate at cost the cavern required. Its great depth would provide immense quantities of protection . . . but even so, would not be capable of keeping out the cosmic ray neutrinos. Nor would it screen out yet a further source of noise: the faint, residual radioactivity of rock, of the solid body of the Earth itself. All such intruding particles, however, would possess a characteristic identifying signature: They would arrive one at a time. The proton, in contrast, was expected to decay into two particles traveling in opposite directions. It would produce two light bursts back to back, as opposed to one. So to separate the signal from the noise, the IMB people would have to determine not simply the presence of each light burst, but its precise pattern. Each event would trigger not just one but an entire sequence of phototubes. The pattern of receptions would have to be recorded on a computer, processed, and later displayed in a complex array, color-coded to indicate which tubes fired first, which second, and the like. The ultimate design called for the ability to distinguish events 0.00000001 of a second apart.

The cavern housing the experiment is the size of an apartment building, sixty feet wide by seventy-five feet high. It is lined with plastic sheeting and contains eight thousand tons of exceedingly pure water. Suspended within, dangling from a maze of wires, hang the phototubes. From time to time a scuba diver descends, adjusting them. The water is so clear that sometimes it is difficult to see. The cavern seems empty, the scuba diver suspended weightlessly in thin air.

The diver completes his work and retreats. In the enormous room the lights go out. A stillness falls, and blackness, the oppressive stygian blackness of the cave. A blackness relieved from time to time by the smallest, the most infinitesimal of flickers of light . . .

Buddha said it: "All composite things decay. Strive diligently."

The experiment is expensive. Its price tag is measured in millions of dollars, and no one would claim the IMB people have been overspending. Indeed, the prevailing opinion is that they have been cutting corners, and rather regularly at that. Nor is theirs the only effort under way. At the time of this writing a Harvard-Purdue-Wisconsin group is operating in a Utah silver mine, and a Japanese group in a mine outside Tokyo.

No one would spend such sums, go to such lengths, without good justification. It would be like drilling for oil in the backyard on the mere shadow of a hope. But there are reasons for believing the proton to decay, and they grow out of some of the most exciting scientific ideas to have been developed in our time. Furthermore these ideas do not just point to proton decay, but also appear to show the way to an understanding of that mysterious process whereby baryon number was created out of nothing in the early universe. And if we are lucky they may resolve the horizon, smoothness, and flatness problems as well. The IMB experiment is one side of a coin. The other side is grand unification.

The thrust of science is twofold. On the one hand it makes the world more complicated. Naturalists scour the continents and plumb the ocean depths, uncovering new and previously unsuspected forms of life; astronomers scan the skies and discover exotic types of stars; physicists probe the atom and find within a wealth of structure. What once was simple is found to be complex; what once was complex becomes yet more so. The textbooks grow thicker: Bookshelves groan beneath their weight. And yet coexisting with this thrust, and operating at the very same time, is its precise opposite. The world grows simpler. Vast overarching principles are discovered, immense generalizations. The near-infinite variety of life forms is comprehended in terms of the universal biochemistry of proteins, amino acids, and DNA; the multiplicity of stellar types in terms of stellar evolution; atomic structure in terms of the laws of quantum theory. The vast wealth of data is simplified into a few broad categories, and unification is achieved.

Newton did that. The child asks why the Moon does not fall down. The wise parent chuckles indulgently at the question—but it took the genius of Newton to realize the Moon *was* falling; falling perpetually, falling incessantly, plummeting in a gigantic swooping arc entirely about the Earth . . . falling as the apple falls. The story that Newton thought of gravitation upon being hit on the head by an apple is apocryphal. What is not apocryphal is that he discovered an underlying identity between these two phenomena. He unified them.

The nineteenth century witnessed the discovery that gases in the laboratory, when heated, glow with a characteristic set of colors—a characteristic spectrum. In 1859 the physicist Gustav Kirchhoff employed this principle to interpret the spectrum of the Sun as evidence of the presence of certain gases there. Here too was unification: Kirchhoff gathered into one broad sweep of understanding the glimmer of a star and the shining of a flame, the contents of a flask upon the laboratory bench and the composition of the heavens.

Most compelling of all in terms of sheer mathematical beauty was the unification achieved by James Clerk Maxwell among electricity, magnetism, and light. In Maxwell's time these were thought to be separate aspects of nature. And why should it not be so? Magnetism was the horseshoe magnet and the mariner's compass, electricity the gently humming motor and the lightning bolt. And as for light, it was no more than pure, insubstantial radiance. What relation could there be among three such wildly disparate things? And at a deeper level, Maxwell knew, the differences were greater still. Nevertheless, in an extraordinary act of unification, he discovered that at heart all were but differing facets of the same underlying unity, the electromagnetic field.

By the mid-1960's the program of unification had progressed so far that all the myriad phenomena of the physical world had been reduced to the operation of a mere four fundamental forces. Most powerful of these is *the strong force*. Discussed in Chapter 7, it is this that binds the atomic nucleus together, fuels the nuclear reactor as well as the atom and hydrogen bombs, and powers the shining of the stars. Next comes *the weak force,* also discussed in Chapter 7, which is responsible for certain transformations of subatomic particles and plays a role in radioactivity. But despite its name, this is not the weakest of the four. Weaker still is Maxwell's *electromagnetic force,* and weakest of them all is *gravity.* In the late 1960's and early 1970's a further unification was achieved, with the development of a theory combining the electromagnetic and weak forces into one, the so-called *electroweak force.*

Then, in 1974, the Harvard University physicists Howard Georgi and Sheldon Glashow took yet a further step, and created a theory uniting the electroweak and strong forces. Three of the four fundamental forces of nature are encompassed in this grand unified theory, or GUT for short. Georgi and Glashow's work represents an immense generalization, breathtaking in its sweep, and it is one of the most important steps forward in the physics of our time. The grand unified theory gathers into one vast pattern the squeaking of chalk on a blackboard and the shining of the stars, radioactivity and the internal combustion engine, the compass needle and the nuclear reactor. Of all the phenomena of physics, only that of weight—gravity—is excluded from its scope.

The theory makes a number of remarkable predictions. In particular, it predicts the proton should decay.

The proton is one of the building blocks of nature: a sub-atomic particle out of which all matter is constructed. For generations physicists referred to it as an elementary particle. But it is not an elementary particle. In recent years the realization has slowly been dawning that the real elementary particles, the ultimate building blocks of the world, are the quarks. There are a number of types, including the so-called u or "up" quark and the d or "down." Just as the atomic nucleus is a composite structure made of neutrons and protons, so the neutron and proton themselves are made of u and d quarks. The proton is uud (Figure 38).

Proton:

Figure 38

According to the grand unified theory, two u quarks can occasionally combine to produce a new particle, known as the X. This X particle, in turn, survives only briefly before decaying. On the one hand, it can decay right back into the two u's from which it came. However it is also capable of doing something else: of decaying into an antielectron and an anti-d quark. The process is schematically diagramed in Figure 39, in which bars denote antiparticles.

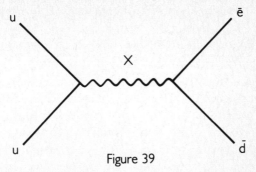

Figure 39

This figure, however, is incomplete. It does not show the d quark already present within the proton. In Figure 40 the error is remedied. From this diagram a new wrinkle emerges: The original d quark, in combination with the anti-d produced by X decay, combines to form yet another particle, known as the pi meson. The proton has spontaneously transmuted itself into a pi meson and an antielectron (neither of which are baryons). They speed away in opposite directions and, if the decay occurs in a medium such as water, emit faint flashes of light as they traverse it. It is this light that the IMB experiment is designed to detect.

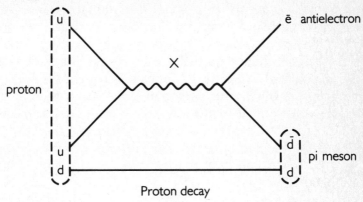

Proton decay

Figure 40

But the process is not yet complete. The pi meson is unstable, and very rapidly it too decays. It decays into light—light of an exceedingly short wavelength to be sure, gamma radiation, but light all the same. And as for the antielectron, it soon encounters an ordinary electron and annihilates against it—also into light. The net result of all these reactions is the transformation of the proton into pure radiance.

But protons are everywhere. There is not a single object on Earth without them. All things are made of atoms, and atoms contain protons. Thus the remarkable prediction of grand unification is that all things glisten. That ashtray on the table over there, for instance—a ghostly radiance emanates from within it, a delicate light streaming outward in all directions. It glows like a light bulb. This radiance fills the room; by its presence it illuminates everything. As does the cigarette butt inside the ashtray and the table on which it stands. Indeed, the very room itself is faintly luminescent. Light streams up from the floor and down from the ceiling.

Tiny cracks appear. Little vacancies pop into being here and there. As time passes these vacancies accumulate, join together into larger structural deformities. Substances once solid begin to lose their coherence. The ashtray is in the process of dissolving. It is dissolving into light. Each time a burst is emitted, another bit of its structure has evaporated.

This emission cannot be suppressed. Wrapping the ashtray in a blanket would do no good—among other things, the blanket itself is faintly luminous. But perhaps the most remarkable thing of all about the emission is that it is totally independent of the body that is doing the emitting. Solid or liquid or gas, it makes no difference; all radiate at the same rate. Green things and blue, flat ones and rough, the corpse in the coffin and the living person it once was—these conditions are irrelevant. It is not the nature of the body that is in question, nor the nature of its composition. It the nature of matter itself, of pure, primordial substance. Substance dissolves into light. The only reason we do not notice it is the slowness of the decay.

It is common enough as a metaphor, as a religious symbol. Dreamers have talked this way for ages. But according to grand unification it is not a dream. It is the literal and exact truth. We are composed of incipient radiance.

*　　*　　*

Grand unification is nothing if not venturesome. In every respect it represents an extrapolation of a gigantic magnitude from the known into the unknown. The lifetimes of most unstable particles are measured in fractions of a second. The theory's prediction for the lifetime of the proton, on the other hand, works out to a mind-boggling 200,000,000,000,000,000,000,000,000,000,000 years. The grand unified theory is equally dramatic in other ways. Elementary particle physics is the study of matter at the finest resolution possible. Current technology is capable of probing the proton to a resolution of roughly one thousandth of its diameter. Grand unification, on the other hand, blithely deals with distance scales some 10,000,000,000,000 times smaller; that is the ratio between the distance to the Sun and the size of an acorn. And finally, the X particle, which plays so crucial a role in proton decay, is thought be similarly massive as compared to other subatomic particles.

To produce a particle and study it in the laboratory requires energy: The more massive the particle and the finer the resolution at which one wishes to study it, the more energy is required. For this reason particle physics is high-energy physics, and the experimental tool of the field is the particle accelerator. At present the most powerful accelerators we possess are those at Fermilab in Illinois and CERN in Switzerland. These machines are immense, measuring thousands of feet in diameter. But in 1983 a special High Energy Physics Advisory Panel recommended the construction of a truly enormous instrument, the Superconducting Super Collider. The Collider as the panel envisages it would be an entirely unprecedented machine, carrying current technology far beyond proven ground. The energy it could attain and the particle types it could produce overwhelmingly surpass those of Fermilab and CERN. These would be rendered obsolete at a stroke. The Collider would be the size of a city, require a decade for its construction, and carry a price tag measured in the billions of dollars. No purely scientific effort of any nature whatever has come close to it in scale.

But even so gigantic a machine would come nowhere near grand unification energies. It would be utterly incapable of producing the X. Indeed, so immense is that particle's mass that an accelerator capable of creating it would quite literally stretch from here to the closest star. It is through the appearance of the X within a proton that that particle decays. It is by virtue of the X that the law of conservation of baryon number is violated. Indeed, the X is central to the entire grand unified theory. But there is not the slightest hope of ever creating one in the laboratory. We do not have enough energy.

But the Big Bang had enough energy.

As emphasized in the previous chapter, light of high energy is capable of initiating pair-creation reactions. Early in its history the universe was therefore flooded with particles and antiparticles. As for *which* particles were produced by these reactions, the answer depends on the energy of the light, which is to say on the temperature of the universe. The more massive the particle the higher the temperature required, and the earlier in the history of the cosmos it existed.

The state represented in Figure 35 (page 146), the state containing more particles than antiparticles, specifically included protons, neutrons, and electrons, the building blocks of which the present cosmos is constructed. Pairs of these particles flooded the universe early in its history. But at yet earlier times the temperature was higher and the X too must have been present. The temperature in that primordial moment was an extraordinary 1,000,000,000,000,000,000,000,000,000 degrees. Such a temperature is incomparably greater than anything we are capable of producing in the laboratory. It is greater than anything nature is capable of producing as well, other than in the Big Bang itself; not in a star, not in the vicinity of a black hole, not in any of the mighty explosions known to astrophysics is it found. But at the moment of creation that temperature was reached, and it was surpassed for the barest fraction of a second—for the first 0.00000000000000000000000000000001 of a second. In that climactic instant the cosmos was flooded with pairs of the mysterious X. It was the time of grand unification.

And during that time the law of conservation of baryon number was violated—not just rarely, not just once in an inconceivable age as with proton decay, but powerfully. It was then that baryosynthesis occurred. It was then that the prevalence of matter over antimatter in the universe was established. Equal numbers of X and anti-X particles populated the universe. They decayed, and unlike every other particle known to physics they did so asymmetrically, into matter more often than into antimatter. As a consequence of these decays, the baryon number of the cosmos, initially zero, began increasing. And then, the barest fraction of a second later, the temperature had dropped and the X's had vanished. The moment of grand unification had passed.

Grand unification is complicated, and the details are much in doubt. Calculations are fraught with difficulty and subject to great uncertainty. "To get a [specific answer] out of this morass is akin to pulling a rabbit out of a soup of amino acids," one worker in the field has quipped. Nevertheless it seems clear that the scenario might work. No one would claim a definitive solution to the matter-antimatter problem. But everyone would agree that the possibility of reaching a solution exists. For the first time in the history of modern science, an answer may be within our grasp to the age-old philosophical question of why there is something rather than nothing at all.

The questions raised by the horizon and smoothness problems are clearly related. The first is how the temperatures of two widely separated regions could be so close, the second how the distribution of matter within them could be so alike. Each underscores the surprising degree of homogeneity of the cosmos. Furthermore, as emphasized in the previous chapter, a departure from each type of homogeneity would imply a departure from the other.

But as with the matter-antimatter problem, it is worth pausing to ask why the horizon and smoothness problems should be thought of as problems at all. Why shouldn't the cosmos be uniform? What is so surprising about its homogeneity? After all, one region of the Pacific Ocean looks pretty much like every other. The atmosphere here is not so very different from the atmosphere there.

But in reality one region of the Pacific Ocean does not look pretty much like every other, and the atmosphere is not so uniform as one might think. *Here* might lie in the tropics, *there* just off the Antarctic coast. In the first region we find coral reefs, warm and balmy breezes, and outrigger canoes; in the second the penguin, the iceberg, and the bitter blast. It is only if the regions under discussion lie close together that they resemble one another. Physical conditions at one point in the ocean are highly similar to those a foot or so away. The weather at city hall usually resembles that uptown.

The universe is remarkable because its structure is precisely the opposite. Only when studied over the very greatest distances does its uniformity become evident. Regions relatively close together, on the other hand, differ radically: The first might lie in the heart of the Sun, the second somewhere out in interplanetary space. These small-scale departures from homogeneity are no mystery; they arise from gravitation's tendency to magnify the slightest irregularity. But the overall homogeneity is exceedingly difficult to comprehend.

One might think the Big Bang theory supplies the answer. After all, according to it the present state of the universe arose from a more compact one—from the Big Bang. Was not the cosmos sufficiently compressed back then to achieve homogeneity? It was not. Return to Figure 31 (page 142), depicting two regions of exactly the same temperature lying a full 34 billion light-years apart. They are not that far apart now. They were that far apart during the Big Bang, shortly after the creation of the universe. Even then they lay at enormous distances from one another. And no matter how much further back in time one projects, the same difficulty emerges: No matter how compact it was, the cosmos was never compact enough for a long enough time to achieve homogeneity.

In 1981 the physicist Alan Guth, then at the Stanford Linear Accelerator Center, proposed a remarkable and ingenious solution to the enigma. Guth stood the horizon and smoothness problems on their heads. They were not problems at all, he decided. They were clues. They were clues to the fact that at some point in the past the universe *had* been small enough to achieve homogeneity.

Guth was saying that, early in its history, the cosmos had been far smaller than even the Big Bang theory claimed. It had been ultra-compressed. While in that state, normal processes had operated to bring about its homogeneity. About those processes there was little mystery. Nor was there any difficulty in postulating the initial, highly compact state. In fact the real question raised by Guth's proposal lay elsewhere. It had to do with expansion. If shortly after its creation the cosmos had been so exceedingly compressed, how did it reach so rapidly the state diagramed in Figure 31? How did those two regions, once literally jammed on top of one another, achieve a separation of 34 billion light-years in so brief a span of time?

They must have reached that separation as the result of an immense outward rush. The cosmos must have expanded— enormously, and in a short amount of time. Guth termed that rush inflation.

Inflation had a second interesting feature as well. It smoothed out whatever inhomogeneities might have survived the initial, ultra-compressed state. Even if the various homogenizing processes had not succeeded in producing a configuration of perfect uniformity, Guth realized, inflation in and of itself would do the trick. A good analogy is that of blowing up a lumpy balloon. Guth imagined a race of intelligent ants dwelling upon the surface of such a balloon, surveying its structure out to some limiting distance with their telescopes. Prior to the balloon's inflation, the ants would easily be capable of detecting its numerous bumps and valleys (Figure 41). After inflation the bumps and valleys would still be present. But they would be distributed over greater distances. The ants, surveying with their telescopes to the same limiting distance, would no longer be capable of detecting them. They would observe the balloon to be smooth (Figure 42).

The endless reaches of interstellar space, the Milky Way Galaxy with its mighty spiral arms, the inconceivable void of intergalactic emptiness, and the immensity of structure of the observable universe—all are nothing compared to the actual scale of the cosmos: a mote, the merest grain of sand in God's immeasurable eye.

*   *   *

GRAND UNIFICATION AND THE INFLATIONARY UNIVERSE

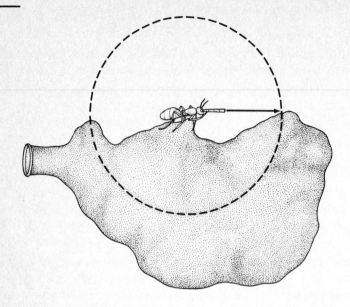

Figure 41

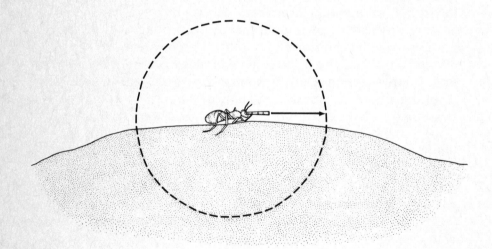

Figure 42

So inflation solves the horizon and the smoothness problems—solves them twice, in fact. Remarkably enough, it solves the flatness problem too, and it does so in a similar way. Recall that the flatness problem refers to the effects of gravitation on the expansion of the cosmos. The question is why, at the moment of creation, the density was so very close to critical—to the value above which the attraction of gravity would have been sufficient to reverse the expansion, instantaneously destroying the cosmos in the fires of a second Big Bang.

Einstein's general relativity is the modern theory of gravitation, and according to Einstein gravitation is a manifestation of geometry. Within relativity, every question one wishes to ask about gravitation and its effects on the universe is recast into a new form concerning the cosmos's geometry. This geometry is not necessarily that of Euclid. Indeed, according to relativity it will be so in one and only one circumstance: if the density of the cosmos is exactly critical. Only then will the geometry of space be that of figures drawn on a plane (Figure 43).

Figure 43

On the other hand, if the overall density deviates from critical, geometry will be distorted into unfamiliar forms. If the density is less than critical, the geometry will be that of figures drawn not on a plane but on saddle, as in Figure 44. A cosmos at greater than critical density, on the other hand, possesses the geometry of the surface of a sphere (Figure 45). And finally, if the universe is clumpy in its distribution of matter, the geometry will be irregular too.

In relativity's language the question is why the geometry of creation was so nearly that of Euclid. Hence the term *flatness problem*. But Guth's inflation supplies the answer. Just as the surface of a lumpy balloon grows smoother as it is inflated, so too it grows more nearly flat. A race of ants upon it would be able to observe with telescopes the curvature prior to inflation, but afterward would be entirely unable to do so. The curvature would extend over such gigantic distances as to be undiscoverable. Their problem would be like

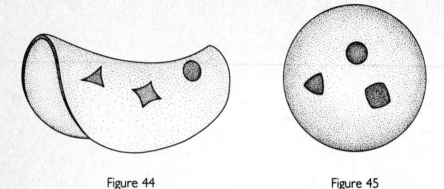

Figure 44   Figure 45

that of attempting to prove the Earth is round by means of observations confined to one's own backyard.

A cosmos ultimately destined to recollapse has the geometry of the surface of a sphere. Gravitation's strength is related to the sharpness of the curvature: strong curvature, strong gravity. A small sphere, on the one hand—a BB—is sharply curved. This corresponds to a powerful gravitational force: Small spheres recollapse quickly. A large sphere, on the other hand—a basketball—possesses a weak curvature and does not recollapse for ages. At the moment of creation the geometry might well have been so sharply curved that the cosmos was doomed to recollapse into universal catastrophe within a mere few minutes. But a fraction of a second later Guth's inflation commenced, and the BB blossomed outward. In an instant, the cosmic lifetime had been extended from minutes to billions of years. So the cosmos escaped an early death.

Guth's inflation gave us time—time for the long, winding course of cosmic history to blossom forth into the phenomenon of life, for evolution to proceed to the development of intelligence, and for humanity to appear on the scene some 17 billion years after that inflation took place. But this immense outward rush did more than stave off recollapse. It also set the stage for life's emergence. Inflation achieved this by solving the horizon and smoothness problems—by smoothing the cosmos into uniform, seamless perfection. Had baryosynthesis not occurred, the cosmos would have contained no matter; had inflation not occurred, that matter would have existed, but it would have remained utterly sterile.

In proposing that the cosmos expanded, Guth was doing nothing new. People had known for decades it was expanding. But according to Einstein's Big Bang theory the rate of this process was relatively sedate: a smooth, not unduly rapid puffing up of a balloon. Guth's proposal was of an expansion over and above that of Einstein. It is difficult to appreciate the enormous, the incomparable magnitude of this inflation. A similar puffing up of a single BB would expand it till it was vaster than the Earth, vaster than the Sun—vaster, indeed, than all the depths of intergalactic space out to the very limits of the observable universe. And furthermore, this extraordinary outward rush Guth claimed to have taken place in a mere 0.00000000000000000000000000000001 of a second.

Such expansion goes far beyond the normal. No ordinary theory of physics would be capable of accounting for such a thing—not even Einstein's. Just as energy is required to blow up a balloon, so energy is required to expand the universe; and just as inflation overwhelmingly surpasses the normal expansion, so too the quantity of energy required to drive inflation. Guth's gigantically accelerated expansion could only have been caused by the sudden appearance of a new form of energy, unprecedented in its magnitude, and incomparably greater than anything previously known to science.

Where did this energy come from? It came from grand unification.

The fundamental postulate of grand unification is that the strong, weak, and electromagnetic forces are different manifestations of the same underlying unity. The present state of the cosmos is such that the unity is hidden: The strong force is far stronger than the weak, which in turn greatly exceeds in strength the electromagnetic. Nor are the differences merely ones of degree, for the three forces act in different ways. But early in the history of the cosmos their unity was revealed: All three were of equal strength and all three behaved the same way. There must have been a transition between the two situations. The grand unified energy appeared, and Guth's inflation occurred, when that transition took place.

Grand unification speaks in terms of symmetry—of the underlying symmetry of the three forces—and in terms of symmetry breaking—of the means whereby this symmetry is hidden at present. A good analogy is that of a ball on top of a hill. As drawn in Figure 46 the situation is perfectly symmetrical—the same in all directions. On the other hand, it is also unstable. Eventually the ball is going to roll off. As to which direction it will roll, this cannot be determined—perhaps to the west, perhaps northeast. But in any event, once the ball has rolled the symmetry has been broken. The configuration is no longer the same in all directions (Figure 47).

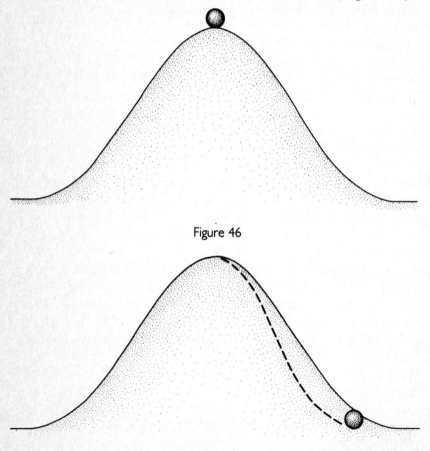

Figure 46

Figure 47

A second analogy is that of air in a box—humid air, air in which some water has been dissolved. The water molecules spread uniformly about the box. Why do they spread so uniformly? Because the situation is symmetrical; they have no reason to congregate in any one place as opposed to any other. But now cool the box. The water vapor condenses. It condenses into rain. If a relatively small number of water molecules have been dissolved, a sufficient quantity will be available to make only a single drop. The raindrop falls downward to the box's floor. Where on the floor will it land? As before, this cannot be determined; it might land close to the right-hand wall, or perhaps somewhere more nearly in front (Figure 48). But again, no matter where the raindrop lands the water molecules will no longer be spread uniformly throughout the box. The symmetry of their distribution in space will have been broken.

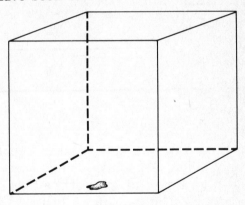

Figure 48

The second analogy has a feature not shared by the first— the breaking of the symmetry is associated with a lowering of the temperature. When hot the configuration is symmetrical, when cold asymmetrical. The same is true of the symmetry envisaged by Georgi and Glashow: It is at high temperatures that unification is achieved. As usual with grand unification, the temperature required to reach this state is far beyond anything that can be attained in the laboratory. But it is not beyond what was once attained in the grandest laboratory of

GRAND UNIFICATION AND THE INFLATIONARY UNIVERSE

them all, the cosmos itself at the moment of creation. In that primordial instant, three of the four basic forces of nature were united into one, and the underlying symmetry of the cosmos was revealed. It was a state of great and seamless perfection. So the book of Genesis speaks of the Garden of Eden, in which the first man and the first woman dwelt before the Fall.

And as mankind fell from perfection, so too the perfect symmetry of grand unification survived for only an instant. The ball rolls from atop the hill; the raindrop forms from vapor. So too, as the universe expanded and cooled, it underwent a transition of profound significance—its symmetry was broken. The grand unified force separated out into the strong and electroweak forces (Figure 49).

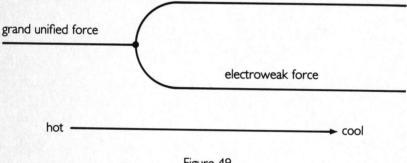

Figure 49

Water, upon cooling, passes through two such changes of state: from vapor to liquid, a transition that occurs at the boiling point, and from liquid to solid at the freezing point. So too with the cosmos. The transition illustrated in Figure 49 was only the first. As the universe continued expanding and cooling, a second critical temperature was reached, and a second change of state occurred in which the electroweak force divided into the electromagnetic and the weak forces (Figure 50).

Nevertheless, just as the hill remains symmetrical even after the ball has rolled, just as the box is symmetrical whether the distribution of water within it is or not, so too the strong, weak, and electromagnetic forces at heart are one. That they do not appear so is the result of a double symmetry breaking that occurred shortly after the creation of the uni-

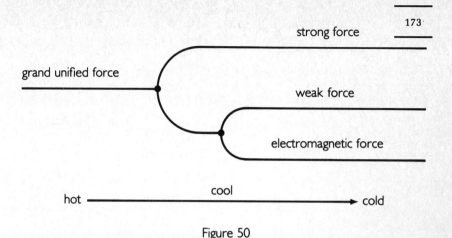

Figure 50

verse, but did not effect their underlying identity. We live in a time of an immense blindness to the unity of things.

The second of these two transitions had little effect on the expansion of the universe. But the first did. It was accompanied by that immense release of energy which drove Guth's inflation.

Return to the analogy of water vapor in a box. The vapor's configuration is not just symmetrical. It is also disordered—the water molecules fly about randomly. The raindrop, on the other hand, is a structure in which each molecule finds its proper place and stays there. Condensation has ordered the molecules' positions in space. The freezing of this drop upon further cooling would order them yet more, into the delicate and complex structure of a snowflake. So too with the molecules' orientation: In the high-temperature vapor state they spin about randomly, while in the low-temperature solid state each is rigidly locked. In every case symmetry is associated with disorder, asymmetry with order. The degree of order can be measured in terms of a so-called order parameter, as in Figure 51. Particle physics adds one more element to the diagram: The disordered state is the grand unified state (Figure 52).

What has this to do with cooling? Why does rain form from vapor, and why do the strong and electroweak forces separate out as the temperature is reduced? The answer involves an analysis in terms of energy. Every system naturally seeks the state of lowest energy. When the system is hot, this

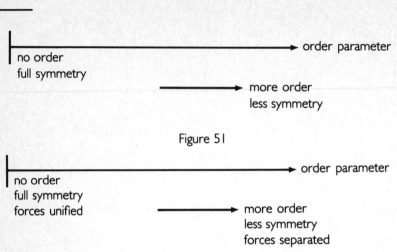

Figure 51

Figure 52

state turns out to be symmetrical, disordered—vapor in the case of water, grand unification in the case of the cosmos. But when the system is cold, the lowest-energy configuration is one of less symmetry and higher order.

In Figure 53 the energy is sketched in terms of the order parameter for the high-temperature case. The sketch looks somewhat like that of a valley floor. Just as a ball placed high on the side of such a valley will quickly roll downward to the center, so the system rapidly reaches the symmetrical state. So hot water boils. But note in this figure that slight indentation high up on the valley slope.

As the temperature is reduced this indentation grows more pronounced. It grows into a second, subsidiary valley perched high above the main one. And as the temperature is reduced yet further, the subsidiary valley moves down toward the level of the first. Finally, at a certain critical temperature the two are equally deep (Figure 54).

Below the critical temperature the situation is reversed. The so-called subsidiary valley has become the main one, and the original symmetrical state is no longer the one of lowest energy. Figure 55 diagrams the new configuration. The ball has hopped over to the new point of lowest elevation. This hop of the ball is what constitutes the change of state. Vapor condenses into liquid, liquid into solid—and forces separate out in the early universe.

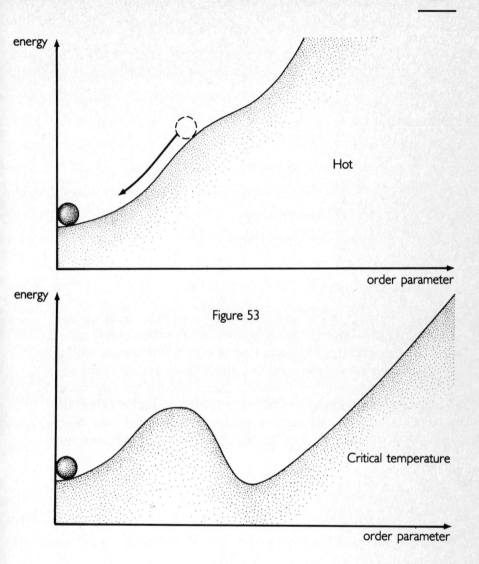

energy

Hot

order parameter

Figure 53

energy

Critical temperature

order parameter

Figure 54

To this picture Guth added one additional element. He asked how the ball could have made that hop. After all, there was a hill in the way. A certain amount of "joggling" was required to push the ball over it. But what if that joggling was not available?

In this case the ball would remain in the original symmetrical state—even at low temperatures (Figure 56). It would be an anomalous configuration, corresponding to

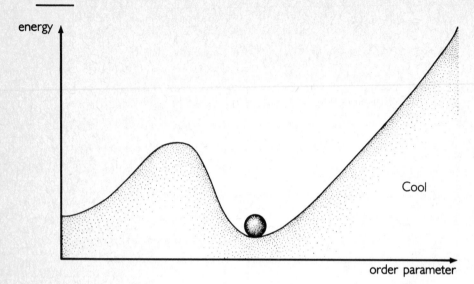

energy

order parameter

Cool

Figure 55

water's remaining vapor below 212 degrees Fahrenheit or liquid below 32 degrees. This is known as the supercooled state.
Remarkably enough, supercooling is not so rare as one might
think. Water vapor very rarely condenses into a liquid exactly at 212 degrees, nor freezes at 32. Rather, it usually waits
until the temperature has fallen somewhat lower, making the
transition only when the molecular arrangement has been
disturbed in some fashion to allow the ball to overcome the
hill.

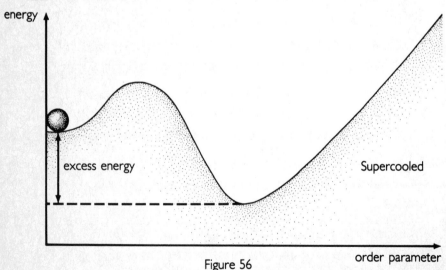

energy

excess energy

Supercooled

order parameter

Figure 56

THE SYMBIOTIC UNIVERSE

But in the case of the cosmos the hill is immense. The universe, Guth decided, would supercool, and it would do so dramatically. Even when the temperature had dropped far below that at which the symmetry-breaking transition should have taken place, it would still remain in the grand unified state. But according to Figure 56 the supercooled state carries an excess energy. For water this excess is not particularly great, but as usual with grand unification the cosmic excess energy was immense, unprecedented in terms of ordinary physics. Furthermore it grew more so as the universe cooled yet further and the supercooling grew more pronounced. And it was this gigantic energy that drove inflation.

Guth's proposal suffered from a flaw. It concerned the specific manner in which the ball eventually surmounted the hill to reach its proper location in the energy diagram and so end inflation. Within his theory the time and the manner in which this occurred differed from point to point within the universe, just as raindrops form from supercooled clouds randomly. The net result would be a balloon inflated by differing amounts in differing regions. The very homogeneity that inflation had been designed to achieve was lost; scientists referred to it as the graceful-exit problem.

Several years later a variant proposal was put forth by A. D. Linde of the Lebedev Physical Institute in Moscow, and independently by Andreas Albrecht and Paul Steinhardt of the University of Pennsylvania. Their so-called new inflationary universe rested on the realization that in certain grand unified theories, the form of the energy diagram would be modified in such a way as to avoid the difficulty facing Guth's scenario. Unfortunately, their model too suffers from certain flaws.

As for the issue of baryosynthesis, detailed calculations have shown that the baryon number of the universe can be accounted for by the creation and decay of the X particle in the Big Bang only if the mass of the X lies within a certain range. If its mass either exceeds or falls below this range, however, the proposed solution will not work. Grand unification does not predict exactly what this mass should be, and as already emphasized, experiments are incapable of measuring it. This solution then, may well be correct—but equally well may be not be.

The minute flickers of light detected by the IMB experiment do not come from proton decay. They come from noise: cosmic rays, radioactivity, and the like. If the theory of Georgi and Glashow were correct, the experiment should have detected one thousand decays in the first year of its operation. It did not detect a single one. At the time of this writing no instance of proton decay has ever been found—not by the IMB experiment, nor by any of the others currently under way.

But there is not one grand unified theory. There are many. Georgi and Glashow's was only the first. Their central accomplishment was to show the way. By now a veritable menagerie of possible theories uniting the strong, weak, and electromagnetic forces have been proposed, and many of them predict that proton decay proceeds so slowly as to be undetectable by current technology. The negative results of the proton decay experiments leave these recent variants untouched. Among these competing theories, only experiment and observation will be capable of deciding which is correct. Unfortunately, however, those of the usual sort are impossible, for particle accelerators capable of reaching grand unified energies will forever remain beyond our reach. The only experiments possible are those searching for proton decay.

No one working in the field today believes these problems are fatal. None are cause for despair. Rather, they are merely hurdles to be overcome, the sort of difficulties people surmount regularly. An immense and heady excitement is in the air, the faint scent of ultimate victory. The hunt is on in earnest for a theory combining gravitation, last of the four fundamental forces of nature, together with the grand unified force into a single seamless unity. People work overtime.

Prior to Georgi and Glashow's discovery, most scientists believed the proton to be absolutely stable. On the other hand, little experimental evidence existed back then to support this belief. By now the situation has reversed: Most investigators believe the proton to decay—but the experiments, if taken at face value, are telling us just the opposite. It might seem a comment on the gullibility of scientists. But it is not.

It is a comment on the allure of unification. Magnificent ideas are in the offing, and profound insights. For the first time in history a modern, scientific theory proclaiming the underlying unity of all things may be within reach. Connections are being drawn between the structure of elementary particles on the one hand and that of the universe on the other, between the ultimate nature of matter and the moment of creation itself.

PART TWO:

MIND

12

The Watchmaker (Reprise)

The fitness of the environment is no mean thing, and it does not refer to us only. Had the cosmos been unsuited to our existence, it would also have been unsuited to that of every other life form—to grasses and trees, to fireflies, bacteria, and whales. It would have been unsuited even to extraterrestrial life forms of the most bizarre sort. Certain biologists have speculated that organisms might exist somewhere in the universe whose biochemistry is based not upon carbon, as is that of Earth-based life, but upon silicon. Gigantic beings may be discovered someday crawling across the surfaces of tiny asteroids, perfectly at home in the vacuum of interplanetary space. Creatures for which we have no name may at this very moment be burrowing contentedly through the icy nucleus of Halley's Comet. Sir Fred Hoyle, whose discovery of the double resonance in red giants was recounted in Chapter 1, has written a science fiction novel in which an interstellar nebula turns out to be alive, its metabolism maintained by an internal network of electric currents—and not just alive, but sentient. No matter: The fitness of the environment is required for organisms even so bizarre as these.

The concept of a universe suited to the requirements of life refers not to this or that organism, but to all organisms—to universal conditions required by every conceivable life form. These requirements are as follows:

Life requires complexity, and except for the case of Hoyle's hypothetical sentient cloud, this means complex elements. But had the red giant stars not managed to perform their remarkable trick, these elements would not exist. At the very same time, though, life requires order, an orderly flow from place to place, as of nerve impulses or blood. But had space fewer than three dimensions that order would be impossible. Life requires warmth—all life, silicon-based monsters included. No creature could flourish at a temperature of absolute zero. Every organism is a factory, and like all factories each requires an outside source of energy. Stars are the source of that energy, but were the strong interaction slightly less strong, no star could shine. The source of energy must last a long time. Evolution requires immense stretches of time in which to operate. It cannot happen overnight. The same is true of those processes by which life, whatever its form might be, arose out of nonlife. But had inflation not solved the flatness problem, the cosmos might have recollapsed, winking out of existence in an instant; had the proton outweighed the neutron, stars could have shone for no more than a mere few hundred years; and had the strong interaction been stronger, their shining might have been violently unstable and prone to the most terrible catastrophes.

Life requires a stable environment. The temperature cannot vary too much. Minor swings from summer to winter are one thing, but radical excursions from the near absolute zero of interstellar space to the vaporizing heat of a star would be unacceptable. This means that the planets upon which life flourishes must move about their stars in circular orbits. But were stars closer together, or had space more than three dimensions, such orbits would be impossible. Even so speculative a beast as a sentient cloud would have to orbit its energy source, and so would be subject to this requirement. *Life requires a structure.* But if the electron charge did not balance that of the proton, no structures could exist. Stars, planets, weird chitinous beings clinging to the surfaces of asteroids—all would have violently exploded. *Life must be made of matter (or antimatter).* But had inflation not smoothed the cosmos into uniformity, its resulting evolution would have led to a state consisting almost exclusively of giant black holes floating in otherwise empty space. Had baryosynthesis not occurred, the equal quantities of matter and antimatter produced in the Big Bang would have annihilated against each other, and the cosmos would consist solely of light.

There's more. It also turns out that were the cosmos unsuited to life, it would not just be unsuited to every form of life. It would also be unsuited everywhere and for all time: on the Earth and on the moons of Jupiter, on distant planets orbiting unknown suns, in the most far-off galaxies, billions of years in the past and billions of years in the future. It would be universally unsuited to life.

Why? Because that unsuitability would be based not on the particular but on the general. The reader will have noticed that by and large this book has made few references to the particulars of biology. Cell structure, chloroplasts and mitochondria, details of the genetic code—these have had no place in the analysis. Rather, the discussion has by and large been based on physics. And this is as it should be, for equally as well as inanimate objects, living beings obey the laws of physics. The environment is fit for life because these laws are.

And these laws are universal. We are assured that life could not flourish at a temperature of absolute zero—assured not because biology is restricted in some fashion but because physics is. Far off in the reaches of space or deep in the heart of the Sun, ages back in the past, everywhere and everywhen, the laws of nature take their old familiar forms, forms that either do or do not allow for life. What we learn in our Earthbound laboratories about this question is valid throughout the universe. What we find today has been valid for all time.

THE WATCHMAKER (REPRISE)

In the last twenty-four hours you have traveled some 12 million miles. You were carried that distance by the rotation of the Milky Way Galaxy; brought along with you was your home, your neighbor's home—and physics labs. If Newton's law of gravitation were a local thing, confined to some region of space a "mere" 12 million miles in extent, you would be floating weightlessly about the living room by now. If the electron charge depended on location you would have burst apart. In the past century the Earth has swung more than 400 *billion* miles through space—and the testimony of history is that the laws of nature have been valid throughout that immense journey. More technically sophisticated measurements, possessing higher precision and extending over greater spans of time, sharpen this conclusion.

The farther into space we peer, the more evidence we find for the universality of natural law. We find double stars orbiting about one another, locked in the same gravitational embrace that binds the Earth to the Sun—and the falling apple in the orchard to the Earth. We find galaxies orbiting one another, their mutual gravitational attraction extending for millions of light-years. The spectrum of light from a star carries information regarding its composition: No matter how far distant, no matter how much hotter or cooler, more or less massive, each star is found to be composed of the same chemical elements we find on Earth. Radio waves from billion-light-year-distant quasars are interpreted using the same physical principles that govern radio waves right here.

Surveying the cosmos to great distances we are peering backward into the past, for light and radio waves require time to bridge the intervening spaces. In this way the astronomer turns archaeologist. A glimmer of light from a star one thousand light-years distant has taken just one thousand years to reach us. Radio emissions from the quasars have been on their way for fully billions of years. Observations of the cosmic background radiation, which was emitted in the Big Bang, probe physical law over the entire history of the universe. And nothing we have ever discovered gives reason to believe that the laws of physics, upon which the habitability of the cosmos rests, have varied in the slightest over the course of its history.

From the distribution of stars in space to the structure of the atomic nucleus, from the nature of electric charge to the chemical properties of water, from stars to the dimensionality of space, and from the Big Bang to the structure of elementary particles and the underlying unity of all things—in every way

> we find all full of wisdom, and harmony
> and beauty; and all this wise selection
> of means, this harmonious combination of
> laws, this beautiful symmetry of
> relations, directed . . . to the
> preservation, the diffusion, the
> well-being of those living things,
> which, though of their nature we know so
> little, we cannot doubt to be the
> worthiest objects of the Creator's care.

The words of William Whewell. What matter that his Bridgewater Treatise was written more than a century ago, that the grand design we see today was hidden from him? His words remain relevant.

But I would not push that relevance too far. Animating the above passage there is an attitude not present in modern scientific writing. There is something more than the excitement of discovery, the thrill of insight. Above all else, in Whewell's words we discern the passionate faith of a man who has seen firsthand God's love for His works.

Earlier I claimed that Whewell, on the verge of discovering the theory of evolution, had failed because he lacked several pieces of the puzzle. That claim was correct insofar as it went. But surely there is more to the story than that. Whewell had his faith—he believed. He did not *want* to find a natural explanation for nature's grand design. What he wanted was his Watchmaker.

Listen to William Paley:

> Under this stupendous Being we live.
> Our happiness, our existence, is in his
> hand. All we expect must come from him.
> Nor ought we to feel our situation
> insecure. In every nature, and in every

THE WATCHMAKER (REPRISE)

portion of nature which we can descry,
we find attention bestowed upon even the
minutest parts. The hinges in the wings
of an earwig, and the joints of its
antennae, are as highly wrought as if
the creator had nothing else to finish.
We see no signs of diminution of care by
multiplicity of objects, or of
distraction of thought by variety. We
have no reason to fear, therefore, our
being forgotten, or overlooked, or
neglected.

With a faith like that, who needs anything so paltry as a great scientific discovery?

My task until now in this book has been to assemble a list—a list of anthropic coincidences, of misfired rifles in a firing squad. Other such rifles are known, but to my mind those I have presented are the most persuasive. My object has been to convince the reader, by the sheer magnitude of accumulating evidence, that we are faced with a mystery of deep significance. The more knowledge we have gained, the more surprising it has become that life exists in the universe.

So how did we get here? It cannot be a matter of life seeking out the appropriate location in which to flourish, for the evidence refers not merely to the Earth but to the cosmos as a whole—to all reality. It cannot be a matter of evolution surmounting obstacles strewn in its path by the environment, for it is the fitness of the environment itself that is at issue. Neither of these suffice to explain how the rules of the game, the laws of nature, came to be so anomalously suited to the requirements of life. "How did it happen that, with what seem to be so many other options, our universe came out just as it did?" George Wald has asked. "From our own self-centered point of view, that is the best way to make a universe: But what I want to know is, how did the universe find that out?"

Paley and Whewell, of course, had an answer to that question.

Is it possible that they were right? I have spent much effort in demolishing the Bridgewater Treatises' outmoded arguments. No matter: For all their errors, were they on the correct track after all? It would not be the first time sound insight was won from insufficient evidence. More than a century after their passing, have we found at last their Watchmaker—not hiding in the petals of a flower, not lodged in the optical perfection of the eye, but ensconced within the very laws of nature themselves? Was it God Himself who crafted those laws so precisely for our benefit? Using the cold, abstract methods of modern science, have we succeeded in proving the existence of a Supreme Being?

You, the reader of this book, will have to give your own answer to that question. Furthermore, no matter what your choice I have no way of proving you right or wrong. Expertise is not relevant here. Nothing science has taught us can help, for a question like this cannot be answered by reference to any facts. The answer can only come from the realm of opinion and belief.

Paley and Whewell had their belief and you will have yours. But I also have mine. My belief is that the discoveries of science have nothing whatsoever to do with religion. I would argue that any attempt to use science to demonstrate the truth of faith is bound to fail. I say this because science and religion exist in entirely different realms. They have no relevance to one another.

We cannot explain by reference to God the fitness of the cosmos for life. I'll go further: We cannot explain *anything* in this way. It is no more than a confession of ignorance to give a religious answer to a question pertaining to matters of fact. The lesson of history is plain on this matter: Time and time again people have sought to account for some mystery by reference to God's wishes, and time and time again a purely natural explanation has ultimately been found. The story is told that the French scientist Laplace submitted a copy of one of his monumental treatises to Napoleon, and that the emperor, upon leafing through its no doubt daunting pages, commented that he found within them no mention of God. "Sire," Laplace is said to have loftily replied, "I have no need of such hypotheses." To argue that the laws of nature are so remarkably suited to the needs of life because God wanted things that way tells us only about the present paltry state of our knowledge. It is merely a way of saying that we cannot think of any other reason. What difference would there be between that and claiming the Sun rises each day because God wishes it to?

The account offered by science of the rising of the Sun is that the Earth is round and that it spins upon its axis. An account like that tells a story, and the story is one we can comprehend. It gives us insight. It gives us understanding. To say on the other hand that the Sun rises because God wishes it to is a statement of a different logical type. Such a statement is not really an explanation at all. It is rather an affirmation of faith.

The scientific answer to any question invariably turns out to have the remarkable property of calling forth yet more questions. Once we accept the scientific explanation for the coming of the dawn, we are immediately struck by a host of new problems. If the Earth is round, why don't we fall off the underside? Why is it spinning rather than stationary? Why once every twenty-four hours as opposed to twenty-three or twenty-five? None of these problems even existed prior to that explanation. *Science expands, rather than shrinks, the list of unsolved mysteries.* A religious account of the rising of the Sun, on the other hand, has just the opposite effect. It halts the insistent, restless drive. It cools the fevered brow.

A further difference between a religious and a scientific account is that the latter is falsifiable. It is capable of being disproved. One can, after all, find out whether the Earth is round or not—by rocketing off into space to have a look, if by no other means. If it were flat, surely the astronauts would have noticed by now. Similarly, the Foucault pendulum can be used to test the rotation of the Earth. Such a pendulum can be seen in any science museum, and if observed long enough will be found to shift slowly the plane in which it swings. That shift occurs because the planet is rotating beneath the pendulum; if it did not rotate, the pendulum would not shift. The truths of religion, on the other hand, cannot be tested in this way. No amount of evidence would suffice to show that it is not God's love that brings forth the dawn. No experiment conducted in some laboratory could ever prove Him absent.

By the very nature of the statements it makes, religion is insulated from fact. The problems of religion are of an entirely different sort than those of science. They are spiritual, not empirical. If the challenge to the scientist is to build a better experiment, the challenge to the believer is to call forth yet deeper faith.

Religious statements are never tentative. They are absolute. No text I have ever read has affirmed that God *probably* loves the world, or *possibly* is omniscient. But if the truths of faith are subject to no limitations, the truths of science are invariably subject to limitations. To one degree or another, every statement made by science is tentative. Science is the art of doubt.

This may seem strange to the layman. Seen from the outside, science often seems to be the ultimate Big List of Facts. Scientists seem to be incessantly pontificating, and all too sure of themselves. But such an impression, while perfectly accurate so far as it goes, derives from only part of science— the well-established part. Anyone wishing to challenge the scientific account of why dawn comes is going to have a tough battle on his hands. Anyone claiming the Earth is flat will be met with similar derision. But in more advanced realms, at the very frontiers of knowledge, such certainty evaporates away. There one finds only varying levels of tentativeness.

Nowhere is this more evident than in the arguments summarized in this book. For purposes of clarity (not to mention persuasiveness) I have omitted a number of subtleties surrounding these arguments, subtleties that render some of them problematic. Now is the time openly to acknowledge their tentative nature. I have already emphasized the tentative nature of one of the arguments concerning the shining of the Sun, and of the entire program of grand unification and the inflationary universe. The claim that brain function is impossible in two dimensions might also be challenged. It may be possible to circumvent the resulting scrambling of communication pathways. One way would be for each neuron to transmit its signals only at certain predetermined times.

Those neurons for which the signals are destined could be biologically programmed to be receptive to incoming signals at just these times; at all other times, however, during which unwanted signals would be impinging, they would be unreceptive. Whether this method could actually be made to work in a biological setting is a delicate question. But the possibility exists that it might.

I am not entirely satisfied with discussions that have been given of water and its life-giving properties. Henderson's arguments do not strike me as being all that airtight. It seems possible that life might flourish even if water were a more normal liquid. Were ice to sink, would the fact really be of such cosmic significance? It is a geographical accident that most inhabitants of this planet live far from the equator and so are subject to severe winter temperatures. But enormous tracts of land in Africa and South America never freeze, not to mention vast stretches of ocean. As for the moderating effect of water on climate, organisms after all are pretty hardy things. Why could they not survive harsh winters—by hibernation, if by no other means? Furthermore, one can imagine planets in other solar systems with elliptical orbits about their suns, orbits that would carry them closest to the warming fire each winter, farthest each summer. Seasons on such planets would be exceedingly mild with no assistance from water. Whether organisms could function were H_2O not so powerful a solvent is a question for a biologist, not a physicist, but I see no fundamental reason why they could not.

These uncertainties do not strike me as being very serious. None is sufficiently fatal, I feel, as would lead us to abandon the entire concept of the fitness of the environment. Rather they are the sorts of ambiguities that underlie every scientific effort. Many are the great advances in knowledge that have been made in the face of equally flimsy evidence. But I would caution the reader that my feeling of confidence here is just that: a feeling, a gut-level judgment based on experience and intuition.

My counsel to all religious people is not to adopt the arguments presented in this book as evidence for the existence of their god. A church built on the evidence of science is a church built on an insecure foundation. A church built on *any* empirical evidence is built on an insecure foundation, for empirical questions can never be answered with that utter certainty required for religious belief. Only the believer himself is capable of supplying such certainty.

The converse is also true. If the discoveries of science become imbued with religious significance, it is not just the church that is placed in a dangerous position. Science is too. Any church that adopts as one of its tenets a scientific discovery is automatically impelled to defend that discovery against all contenders. The priest becomes a biologist, the rabbi a physicist. The mild skepticism I have just expressed would immediately brand me as a heretic. Committees would be set up to oversee the program of each research laboratory.

In his *Dialogue Concerning the Two Chief World Systems*, Galileo pleaded with the church not to condemn as heretical something that might ultimately turn out to be true. His fate, and those of Copernicus and Giordano Bruno, are examples of what can go wrong when the church sets itself up as arbiter of scientific truth. Every religious person is, or ought to be, ashamed of the role religion played in the scientific revolution. But such are the invariable consequences of any mingling between the religious and the scientific spheres of life.

The religions of the world have been the wellsprings of some of the greatest triumphs of humanity: from the deepest insights into the human condition to moments of the most perfect spirituality, from the noblest acts of self-sacrifice to the most profound works of art. These are the proper realms of faith, and it need not be ashamed of them. But I would argue that faith should stay where it belongs.

Religion is intimately connected with matters that are ultimately human. But the fitness of the environment is not. Nothing that I am saying should be taken as implying that it was us the cosmos conspired to bring into being. The conditions required for humanity to exist are also the conditions required for every other organism to exist. Some fat and happy barnacle, contentedly clinging to a rock in a tide pool, could well wonder how the universe came to be so precisely suited to its needs. That barnacle could ask how the ocean with all its life-giving properties had come into being. It could expound upon the rock to which it clings, so providentially provided for its support. The creature could construct a telescope with which to peer into heaven, an elementary particle accelerator with which to probe the secrets of the atom. It could marvel upon the resonance at work in the red giant stars, without which neither it nor any other crustacean could have come into being. In its mind's eye it could peer back toward the instant of creation itself, and expound on how inflation set the stage for the appearance of barnacles on a tiny blue-white dust mote orbiting about an insignificant sun 17 billion years in the future.

Evil has always been a conundrum for religious thinkers. How an omniscient and a benevolent God could allow it to exist is a problem no religion has put to rest. But evil is not a problem within the concept of a universe fitted for life. Nothing in that concept should be taken as proving that we should not suffer. There is no conflict with the overall fitness of the universe if an individual happens to lead a life of unrelieved misery and pain. It would not even be a conflict if a thermonuclear war were to break forth, eradicating every last vestige of life upon the Earth.

The religious view, I suspect, would tend to celebrate life's appearance upon the cosmic scene. It would uphold the fitness of the universe as an example of God's goodness. But was it really such a good thing that the universe conspired to bring forth life? Why be thankful for the fact? We are accustomed to believing that life is good and nonlife bad, but that is only because of our built-in will to survive. And this, in turn, exists within us for evolutionary reasons: Any species lacking that will would have been doomed to early extinction.

But if we were capable of backing away from our biologically ingrained urge to live, we might well conclude that the appearance of life within the cosmos was no cause for rejoicing. The dog writhing in the gutter, its back broken by a passing car, knows what it is to be alive. So too with the aged elk of the far north woods, slowly dying in the bitter cold of winter. The asphalt upon which the dog lies knows no pain. The snow upon which the elk has collapsed knows not the cold. But living beings do.

Are you alive? Then you can feel pain. Are you conscious? Then you can feel more pain. The lower animals, at least, appear not to be subject to mental anguish. Given the bare necessities of survival, they seem content. But we humans suffer in more sophisticated ways. Perhaps we even suffer more than the dumb animals. No one among us has led a life free from grief, loneliness, and despair. The more that cosmic evolution has progressed, the more suffering has come forth in the world.

There may be no cause for thankfulness in the fact that against all odds the cosmos succeeded in bringing forth life. It may be only the truth.

THE WATCHMAKER (REPRISE)

13

Experimental Metaphysics

For all of the preceding reasons I cannot accept the notion that it was God Himself who so carefully crafted the cosmos in order that it might bring forth life. I reject the supernatural. But in that case, how to account for the habitability of the cosmos? If this is the best way to make a universe, how did the universe find that out?

It may not be possible to answer that question right now. Some mysteries simply cannot be resolved until our knowledge has progressed in ways unseen at present. This may be one such. It may be that all that is left for us is to wait.

Possibly so. But in the remainder of this book I wish to explore a tentative reply to that question, a reply suggested by the great revolution in thought that is quantum mechanics. Quantum mechanics was developed in the first three decades of this century as a theory of atomic structure. The problem of life in the universe could not have been more distant from the minds of its originators. Nevertheless, insights from their theory may have great relevance to the problem.

It may be that the explanation for nature's extraordinary fitness for life must be sought not in the realm of religion, not in the realm of physics or chemistry or biology, but in an entirely different direction. It is to the nature of being, to the very realm of existence itself, that I wish to turn attention. The subject now is metaphysics. Metaphysics, the study of existence and of the ultimate nature of reality, is usually considered part of philosophy. But quantum mechanics too has something to say about the subject.

The proposal I now wish to explore is that in the fitness of the environment we are witnessing the effects of a gigantic symbiosis at work in the universe. Symbiosis is a mutual interdependency of two organisms, each one required by the other and neither capable of surviving on its own. Numerous examples are known, perhaps the most striking of which is the relationship between cows and bacteria. Remarkably enough, a cow is entirely incapable of digesting its principal food, grass. Rather, this function is carried out by bacteria that dwell in the animal's intestine. No matter how much it ate, without these bacteria the cow would starve; conversely, the bacteria depend for their existence upon the comfortable environment provided by the cow.

But the symbiosis to which I am referring here is different. It is of an entirely new sort. The first partner in this new interdependency is not an organism at all, but rather an inanimate structure. Furthermore, it is a structure not previously suspected as taking part in such relationships: the universe as a whole. The second partner in turn, to which the first is inextricably conjoined, is alive, but it is not any particular organism. It is not even an entire species. Rather the second partner is all organisms—life itself.

The proposal is that life and the universe are melded into an immense symbiotic unity, and that this is so for reasons that are ultimately metaphysical. Why did the cosmos bring forth life? It had to. It had to in order to exist.

Two chapters will be required to develop this thesis.

In studying a subject as intimidating and abstruse as the

very nature of existence, it is best to start slowly. Begin at the beginning: with the simplest, the most elementary thing of all—a single particle, the unit out of which all other things are constructed. To make things specific, the following discussion will refer to the electron, but the choice is not crucial. Any other particle would have done as well: the neutron, the quark. Only when the nature of so simple a thing is well in hand can the jump be made to planets, stars, and the universe as a whole. The mysteries of existence can best be approached by a series of experiments. Five will be recounted in this chapter, and each is concerned with the same question: What is the nature of the existence of the electron?

Free electrons are easy to come by. They are emitted by certain radioactive elements, for example. Another means involves a device at the heart of every television set, the so-called electron gun, which produces a steady stream of them directed toward the viewer and falling upon the TV screen. These are in fact moving quite rapidly, and if the screen were removed, so allowing them to escape, the set would make a perfectly reasonable source for the following experiments. The first goal is to show that *the electron is a particle.*

This is most clearly revealed in a cloud chamber. A cloud chamber is a box containing air—humid air. The air in my lungs is humid; if I breathe upon a mirror, a fog spreads over it. The fog consists of water droplets condensed from my breath. Why does the fog appear there but not within my lungs? Because the mirror is at a chilly room temperature, but the interior of my body is perpetually maintained at a comfortable 98.6 degrees. So too I see my breath outdoors on a winter day. Humidity is water dissolved in air; the condensation of this water into droplets is a somewhat difficult process and needs assistance, in this case the assistance of contact with a colder medium.

There are other ways to help along this condensation. A second is to introduce an impurity into the air. A particle of dust will do the trick. Raindrops form in this manner, about tiny grains, microscopic specks of dirt suspended within a raincloud. And finally, a cloud chamber uses electrons themselves as the impurity triggering the formation of a water droplet.

Place the cloud chamber in the path of the electron gun. Adjust the gun to a very low intensity, turn it on, and wait. Nothing happens at first. But then, suddenly and without the slightest warning, a track appears—a track of water drops, a sequence of them arranged in a straight line and extending outward away from the gun. The track is diagramed schematically in Figure 57; an actual photograph is reproduced in Figure 58.

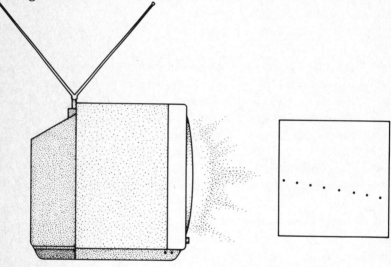

Figure 57

Wait still longer. The track fades as the droplets dissolve again into the air, or are spread about by faint currents within the chamber. Nothing happens for a while. But then a second track appears, pointing in a slightly different direction. (Figure 59). This too dissolves, to be replaced by another and then yet another. The gun seems to be working in bursts.

Now turn up the power. The electron gun still persists in emitting tracks, but it does so more rapidly. The lines of water droplets appear not once every few minutes, but once every few seconds. Still more power. They flow together and form a pattern, a sheath, as in Figure 60.

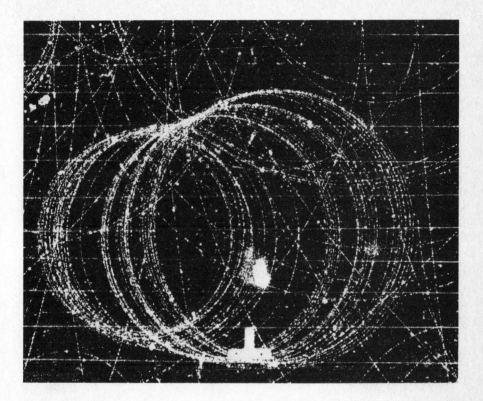

Figure 58: The electron is a particle: tracks in a cloud chamber. The tracks, normally straight, are curved here by an applied magnetic field. Both electrons and antielectrons can be seen, one type curving clockwise and the other counterclockwise. *Omikron/Science Source/Photo Researchers*

It is hard to avoid the impression that the electron gun is emitting particles—tiny bullets fired off into the cloud chamber, triggering the formation of water droplets as they pass. That would certainly explain the tracks. One imagines that what we call the intensity of the gun simply corresponds to the rate of emission of these particles. All well and good: The electron is a particle. But what is the relevance of all this to the nature of existence? What is the problem? The problem comes with the next experiment, which demonstrates that *the electron is not a particle at all.* It is a wave.

EXPERIMENTAL METAPHYSICS

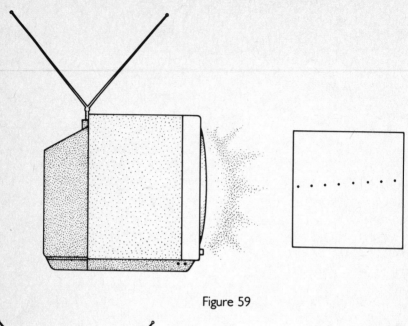

Figure 59

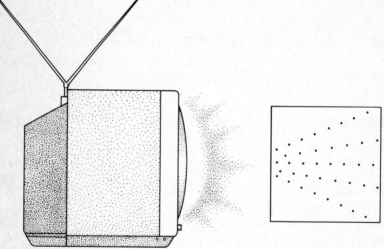

Figure 60

How to demonstrate that something is a wave? One must show it exhibits a property common to waves and waves alone. The property the second experiment employs is that of interference.

THE SYMBIOTIC UNIVERSE

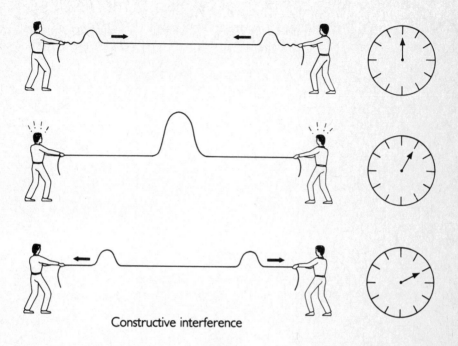

Constructive interference

Figure 61

Figure 61 illustrates the principle. In this figure two people hold a rope stretched tight between them, and both have given the rope a shake upward. In this way, each has launched a wave moving down the rope toward the other. The top part of the figure shows the configuration the instant after the waves have been set off. As for the central part of the figure, it shows the configuration a few seconds later. At this instant the two have met and merged, adding to produce a crest higher than either one alone. The addition is the process of interference—a so-called constructive interference in this case.

Finally, the bottom of Figure 61 illustrates the configuration still later, after the waves have passed each other and continued on their way. Pass on now to Figure 62, which illustrates the opposite case of destructive interference. This is produced by shaking the rope in opposite directions, one person upward and the other downward, as opposed to the two upward shakes of the previous example. Now the waves subtract, completely canceling each other out when they cross. At the instant of merger, the rope lies flat.

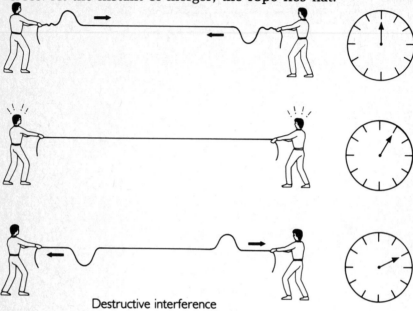

Destructive interference

Figure 62

That is interference. It is not confined to ropes alone; any situation involving waves will exhibit the phenomenon. A stone tossed into a lake will produce an expanding network of ripples. Two stones so tossed will produce a double network, and where the ripples overlap interference will result, as shown in Figure 63. In this situation both types are observed at once: Wherever the wave crests merge, the water tosses up and down quite strongly, while wherever the crest of one merges with the trough of another, the water is perfectly quiescent. The surface of the pond is filled with a pattern of locations alternately exhibiting constructive and destructive interference—an interference pattern.

If electrons are waves, they ought to do the same. The principle of the experiment designed to find out whether they do is illustrated in Figure 64. Here a train of waves is rolling in from the open ocean toward a harbor that is protected by a jetty, the jetty pierced by two gaps through

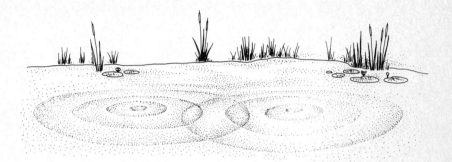

Figure 63

which ships may pass. Of course the waves can pass through those gaps as well, and as they do, a network of expanding semicircular ripples is formed to the right of the barrier. The network is similar to that produced by tossing two stones, and here too the harbor will be filled with the same pattern of constructive and destructive interference. The goal is to find it.

In the present instance the task is trivial—you simply look. But in the case of electrons that method will not do, for electron waves cannot be seen directly. To preserve the analogy, an indirect technique is required. A boat floating in the water will do the trick: Look, not at the waves, but at the boat, bobbing as they pass beneath. The vessel functions as a device for detecting the waves. And finally, to determine the overall pattern of the interference one needs a whole collection of boats, each revealing the state of the water at a different location in the harbor. It will be convenient to string them together in a line, as illustrated in Figure 64.

That is the experimental apparatus: a barrier with two gaps, and behind it a chain of boats. Doing the experiment, one finds that some boats are tossed up and down quite vigorously, others not at all. The degree of bobbing is sketched

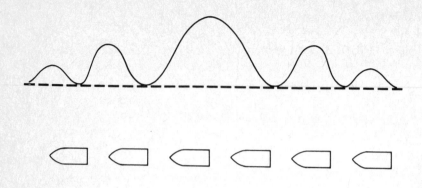

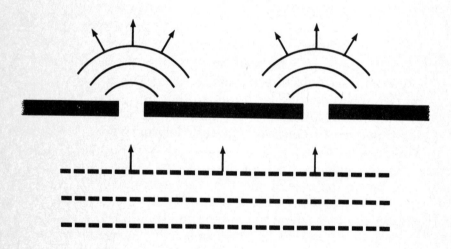

Figure 64

in beside the chain: greatest at its midpoint, diminishing as one scans up or down along it, and exhibiting an alternation between constructive and destructive interference. This right-hand sketch is the characteristic signature for which it is necessary to search. Any phenomenon whose underlying nature is wavelike will exhibit it. If, for instance, a sound wave is sent through a wall pierced by two slits, the sound intensity beyond will show exactly this pattern of variation. And if a beam of electrons is sent through a similar wall, its intensity as it falls upon a screen turns out to show it too.

In the actual experiment, the chain of boats is replaced by a strip of photographic film (which can be exposed just as well by electrons falling upon it as by light). Turn off the overhead lights, close the blinds, and turn on the electron gun. Wait a few minutes and then develop the film. The telltale alternating pattern of light and dark areas will be found. An actual example is shown in Figure 65. Electrons are

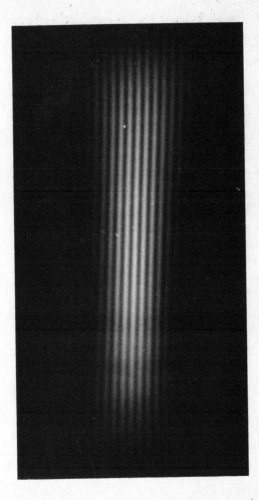

Figure 65: The electron is a wave: an interference pattern produced by passing electrons through two thin slits.   Zeitschrift für Physik

The picture inescapably brought to mind by the first experiment involving the cloud chamber was that of some stream of particles emanating from the electron gun. But that brought to mind by the present experiment involving interference is of something else: of some sort of fluid spilling outward from it, filling all space and capable of supporting waves. Those two pictures are utterly in conflict. One of them must be wrong. But which? Perhaps it is the first. Could it be that our interpretation of that experiment was in error? Does the cloud chamber really prove the electron to be a particle? Or can understanding of its operation be developed based on the notion that the electron is a wave?

Analyze more carefully how the cloud chamber works. What, precisely, triggers the formation of a water droplet? It turns out that the impurity about which the drop forms is an ionized atom. An atom is composed of a nucleus surrounded by electrons; an ion is an atom from which one of the electrons has been dislodged. Evidently the passage of the incoming electron has broken the atom into its component parts. It is easy enough to see how a particle would do that, crashing along like a bull in a china shop. But what about a wave? Is this so difficult a thing for it to achieve? Surely it is not. The atom after all is a composite structure, built of parts, and a wave passing by is guaranteed to shake it a little. A sufficiently intense wave could easily knock loose a few nuts and bolts if the thing was not well constructed. What would be wrong with imagining such a thing?

Nothing at all. It is the *pattern* of broken atoms that cannot be understood on this basis. For after all, each electron leaves not just one ion but many, and all arranged in a perfectly straight line. A wave would not do that at all.

The atoms in the cloud chamber constitute an array of detectors, spread uniformly throughout it and signaling by their breakup the passage of the electron. In terms of the analogy of the water wave, they can be likened to a flotilla of boats. If the electron were a wave, Figure 66 would illustrate the situation as it passed through. It would shake all the boats to the same degree. If the wave were weak, none would be broken apart by its passage—and the chamber would show no tracks.

If the wave were strong they all would be broken, and the chamber would entirely cloud over—and still there would be no tracks. Finally, in the case of a wave of intermediate strength, one would expect either every boat to be broken or none of them, for atoms, after all, are identical to one another in their construction. And even if they were not, those few unusually fragile ones could hardly be expected to lie so precisely along the straight line illustrated in Figure 66. Rather they would be scattered randomly about—and still there would be no tracks. And finally, even if by some extraordinary coincidence they were so arranged, how are we to understand the fact that the next sequence of shatterings, produced by the very same wave, takes place along a different line?

A further difficulty facing the wave theory arises from the fact that, according to it, the intensity of the electron gun is related to the strength of the wave it emits. If so, a weak gun—weak waves—ought to be completely incapable of producing any tracks at all. After all, tiny ripples playing on the surface of the ocean pose no threat to seagoing freighters. On the other hand, the gun certainly does produce tracks in the cloud chamber when adjusted to a low intensity, and indeed it produces them no matter how faint the intensity may be. When one calculates the energy required to ionize an atom and compares this with the wave energy falling upon it from an actual source, the contradiction becomes most evident: The wave does not supply enough energy, and the discrepancy amounts to enormous factors.

No . . . it is no good. Waves don't do such things—ever. The operation of the cloud chamber proves the electron to be a particle. A contradiction is in the works.

That contradiction extends even to the cloud chamber itself, for there are aspects of its operation that cannot be understood if the electron is a particle. The intermittent tracks appearing within it imply that the gun is working in bursts, emitting electrons only sporadically. But the gun is not working in bursts. It is operating perfectly smoothly, operating even during those intervals of time in which the cloud chamber shows no tracks. This can be checked by borrowing one

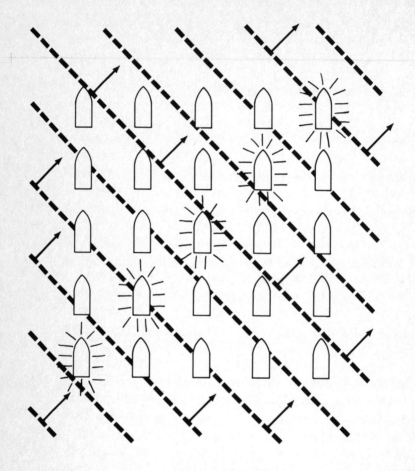

Figure 66

of those meters the electric company uses to monitor home power consumption, and hooking it up between the electric outlet and the gun: The meter will indicate a perfectly steady flow of power.

Things are getting interesting. The electron is a wave and it is not a wave. The electron is not a particle and it is a particle.

*　　*　　*

The next experiment explores further this remarkable interplay. The two-slit apparatus of the second one showed no trace of particles. On the other hand, how could it? Photographic film is not a good device for revealing their presence. Replace that film with a cloud chamber and run the second experiment over again.

Use a whole collection of cloud chambers. A diagram of the experiment in its new form would resemble Figure 64, but with the row of boats replaced by a similar row of cloud chambers, each detecting the arrival of electrons at its particular location. Now there is a means of watching them as they arrive at the detecting apparatus. Once again, turn on the gun and wait. At first nothing happens.

But then a track appears in one of the cloud chambers. Which one? There is no way of predicting in advance. No matter—that is not the point. The point is that the track appears in one chamber, as opposed to several. This is enough to show that a single electron has traversed the experimental apparatus, starting at the gun and ending somewhere in the bank of detectors. The absence of multiple detections is something the wave theory cannot possibly explain. Once again, the electron is a particle.

Keep watching. A second track appears. Where? Again, there is no way of predicting in advance, although most likely it will be found in a chamber other than the first. Better start writing all this down, though. As the chambers keep registering particles, it might be wise to keep a list of which detector records which track.

The longer one waits, the more tracks appear—each in one chamber only, each demonstrating yet again that electrons are particles. On the other hand, as time passes and the list of detections grows longer, a pattern begins to emerge. Looking over the list it becomes evident that certain cloud chambers have recorded more tracks than others. Indeed, some have not revealed a single one, while others have been positively ripping along. Which has recorded the greatest number of electrons? The one at the midpoint of the chain, symmetrically located across from the two slits. Moving up or down the chain from that point, the number of detections drops swiftly to zero, then rises to a maximum, then drops to zero again, then up again to yet another maximum. That's a familiar pattern, of course. Draw a graph showing the total number of electrons detected as a function of position along the chain. . . .

The graph one obtains is identical to that shown in Figure 64. What arrives at a detector is a particle, all right. But what went through those slits was a wave.

"When the going gets tough, the tough get going," runs the saying. Whatever the explanation for this strange behavior, it is not going to be simple. But give it a try. How can the above three experiments be reconciled?

One possibility is to suppose that electrons in large numbers intrude upon one another. Could it be they are particles that suffer a kind of traffic jam, blocking each other's passage as they all attempt to make their way through the slits, and so tricking us into believing them to be a wave? Could such a collection exhibit an interference pattern?

Unfortunately this attempt at an explanation is not going to work. The way to show it is to turn down the intensity of the electron gun. If this is done the resulting interference pattern is found to be entirely unchanged. It just takes longer to build up. But in terms of the particle picture, the gun is now emitting electrons very slowly—once a minute, say. On the other hand, each travels at a high speed and passes through its slit in a fraction of a second. In this case there is no traffic jam at all. Each electron traverses the experimental apparatus alone, separated from all the others . . . and suffers a wave-like interference.

Too bad. But don't give up so easily—try again. How about imagining that electrons are not truly elementary particles? After all, protons are not; they are composed of yet smaller entities, the quarks. Could the same be true of electrons? Perhaps each electron, once emitted by the gun, divides into its constituent parts, the first part to traverse the upper slit and the second part the lower, later to combine into a single unit before arriving at the bank of cloud chambers. This is an intriguing notion, and if true would resolve many contradictions. How to test it? The cloud chambers themselves give a clue. They allow us to see an electron. Add two more chambers to the experiment, one beside each slit as in Figure 67, and so watch each isolated particle as it makes its way through the apparatus. In this way one can determine which slit—maybe even both—it passes through on its journey.

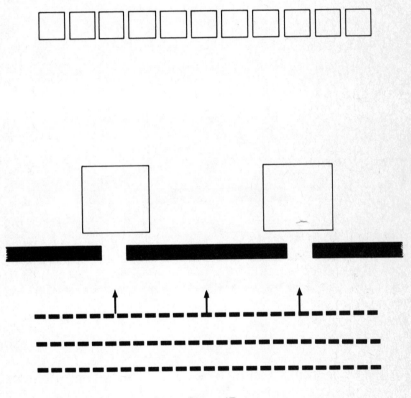

Figure 67

Mount the additional devices behind the slits, lean back, and watch. After a few minutes a track appears. And remarkably enough, it does indeed appear in two cloud chambers—but not the right two. Rather than appearing in the chambers beside the two slits, it appears first in the one by the uppermost slit, and second directly across from that slit in the bank of detectors. The configuration is diagrammed in Figure 68. A clear sense of the electron's path through space is evident. However, because the track did not appear in the detector beside the other slit, this particle, at least, has failed to split in half.

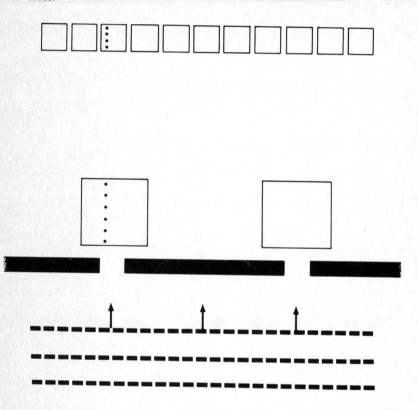

Figure 68

Keep waiting. Different electrons pass through different slits, always accompanied by a track in the corresponding element of the detector bank. Each electron exhibits a well-defined path through space . . . and each passes through one and only one slit. Never does a particle pass through both. Evidently the attempt at resolving the contradiction has failed and we are back to square one: The interference pattern remains incomprehensible.

But wait! Does this experiment show that interference? Run through the list of detections. An astonishing fact emerges. The interference has gone away. The telltale pattern of constructive and destructive interference has entirely disappeared, and in its place is something quite different: two symmetrical bumps directly across from the corresponding slits. This new result is sketched in Figure 69, and it is exactly what would be expected if the electron were purely and simply a particle, and exhibited no wavelike properties at all.

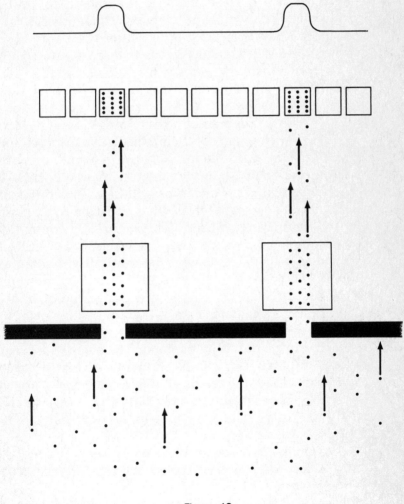

Figure 69

Remove the two cloud chambers from alongside the slits. The interference pattern returns. Put them back again. The interference pattern goes away. If someone is watching, the electron is a particle. If no one watches, it is a wave.

Those electrons from the gun . . . are they watching us?

An insane notion, forced upon us by an incomprehensible sequence of facts. But hold on a moment. Is the notion really so insane? By now it has become clear that the nature of the electron's existence depends upon the experiment that is performed. The thing simply refuses to settle down into anything one can get one's hands upon. Rather, it acts like a recalcitrant boy refusing to satisfy his parents as they demand to know how he managed to get his brand-new clothes so muddy. He has decided to answer each of their questions nonsensically, and he does so by maintaining close attention to their minds, keeping track of their expectations, and giving replies inconsistent with them. Remove now all the anthropomorphic content of this analogy, and ask if such a thing might be possible for electrons within an experimental apparatus. Maybe so—for after all, they must be interacting with the apparatus in its entirety. The apparently contradictory results might be a signal that the electron has indeed been influenced, in a fashion we cannot at present comprehend, by the nature of the experiment we elect to perform. Maybe the film influences it in one way, the cloud chamber in another. In such a manner, concrete form can be given to the unsettling suspicion that we are being watched.

But this too turns out to be no good. The electron is even more subtle than that. This can be shown by one last experiment, and to my mind it is the most bizarre, the most extraordinary and incomprehensible of them all. It is known as the delayed choice experiment, and though easy to describe its implications are revolutionary.

Return to the electron beam directed at the two slits. To detect that beam after it has passed through, the delayed choice experiment uses two different techniques. In both cases it is only concerned with electrons arriving at a particular location—a location of destructive interference. Waves arriving there add up to zero, and to the extent that the electron is a wave, no signal at all is expected at this point.

In the first mode of operation, one simply places a tiny patch of photographic film at such a location. Figure 70 diagrams the configuration. No matter how long one waits before developing the film it is never exposed, for it receives no electrons. So far so good; the result is in conformity with the expectation that the electron is a wave. After a while, though, the experiment grows boring. Try it, then, in the alternative mode depicted in Figure 71. Here the detection apparatus is a bit more complex. It consists of two separate cloud chambers, each arranged to point at a particular slit (this can be done by fixing pipes on their front ends, down which the electrons must pass), and placed at the very same location as was occupied by the patch of film. The upper chamber reveals electrons arriving from the upper slit, the lower chamber from the lower slit.

Sit back and watch. Soon the lower chamber shows a track. As time passes more and more electrons arrive, some in the lower, some in the upper cloud chamber. Now, that is a remarkable result—and it is a new one. With the photographic film in place, electrons did not arrive. But now, with the two chambers there instead, they do. Evidently the electron's mysteries go far beyond the question of whether it is a particle or a wave. They extend to so simple a matter as whether it is or is not arriving at a particular place. Here too, the answer depends on how the arrival is sensed. Walk outside—it is raining? Apparently the answer is not so clear.

So far, though, the experiment has not involved the crucial element of delayed choice, and so has not revealed the electron's most profound mystery. There still remains the possibility that the electron interacts with the detecting apparatus somehow, responding differently to the film on the one hand and the chambers on the other, and so altering its nature. But the next step will reveal this possibility to be untenable. The method is to assemble the experiment in one of its two modes and wait until after the electron has traversed the slits. By that time it will have made up its mind, so to speak, having sensed the choice of detector and settled down into either a wave or a particle nature. It is either going to

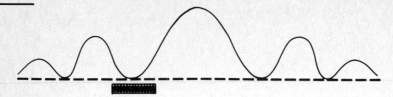

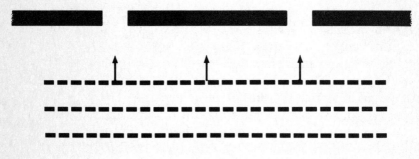

Figure 70

arrive at the detector or it is not. But then, at the very last moment, *change the detector.*

Begin with the photographic film in place. This choice induces the electron to act as a wave. It leaves the gun, traverses both slits, interferes destructively, and so fails to arrive. Now, the electron requires a certain amount of time to do all this. To make vivid the point of the experiment, suppose its progress through the apparatus takes an entire year. The experiment is of cosmic proportions, the gun and slits located far beyond the confines of the solar system while the detecting apparatus sits downstairs in the basement laboratory. On the first day of January, turn on the gun. It is mounted relatively close to the slits, and within an hour the electron has passed through them—and because of the pres-

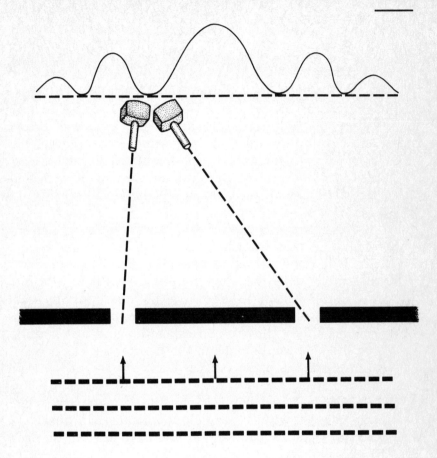

Figure 71

ence of the photographic film, it has passed through both. A pattern of ripples expands outward from the two slits, interfering constructively and destructively at various points in space. The electron is not going to expose the film. On the other hand, not for a year will this become evident.

The weeks roll by and turn into months. Spring arrives. The electron speeds toward the film in the form of waves, and the newspapers are full of nothing but the latest crisis in the Middle East. By summertime the electron has covered half the distance allotted to it, and an oil shortage has raised gasoline prices to an all-time high. By Thanksgiving the expanding network of ripples has created an interference pattern stretching over literally trillions of miles, though its forward edge is still nowhere near the film. It is due to arrive at exactly midnight on New Year's Eve. Polish up those cloud chambers. Get out the clock. Christmas arrives, then the night of the thirty-first. People start handing out party favors and silly hats. It is ten o'clock in the evening. It is eleven forty-five. Everyone is at a party—everyone but you. You are downstairs prowling restlessly about the lab, and those waves are rushing toward it at an enormous rate. The clock ticks. And then, at the very last moment, switch detectors! Leap forward, snatch away the film, and replace it with the two cloud chambers. You have tricked the electron.

Or have you?

The recalcitrant boy replies to his parents' questions in writing. He sees they want him to answer "Yes," and so he scribbles "No" upon a scrap of paper. Folding it, he hands it over. His mother takes the scrap in her hand without looking. And now, at this instant—she reverses her question! She asks just the opposite question. Opening the paper she glances down . . . and throws it away in desperation and rushes from the room. Her husband stoops to the floor and picks it up.

Written upon the paper is "Yes."

At the last moment you leap forward, snatch away the film and replace it with the two cloud chambers. A track immediately appears. By your action you have altered the past one year of history.

This completes the series of experiments probing the nature of the electron. They have been described in terms of electrons from a gun in a TV set, but as already emphasized electrons from radioactive decay behave just the same. As do those in the electric wiring of a house or, indeed, within one's body. The source is not important.

The same enigmas also surround the behavior of protons. The proton is a particle—but on some occasions it too turns into a wave. It has been shown to exhibit the phenomenon of interference if things are properly arranged. Indeed, if the sequence of experiments recounted above is performed on protons, identical results will be obtained. So too with neutrons and, indeed, with the entire world of subatomic particles.

Remarkably enough, the same is true of something quite different: light. Light is a wave, but in certain circumstances it becomes a particle. Our investigation, initially limited to a single species of particle, has turned out to be far broader in scope. An insight has been achieved—an insight into every fundamental entity of which the universe is constructed. But what is that insight?

In surveying the results of these experiments a vast bafflement is likely to seize the reader. They seem utterly insane. Faced with such madness one is almost tempted to say that in some incomprehensible way the very rules of logic must now be abandoned. But it will pay now to pause, and ask just what has led us to so extreme a turn. What is so senseless about those results?

They strike us as senseless because of an assumption we have been making, an assumption so reasonable, so normal, and so universally accepted that we never even bothered to state it. Indeed, most likely we are not even aware of having made it. *We assumed that the electron has a nature to be discovered.*

The electron does not lie. It does not read the experimenter's mind and decide to frustrate him at every turn. It simply does not *exist* in the ordinary sense of the word. And the same is true of every other particle of the subatomic world. The experiments recounted above—and it is worthwhile to emphasize that experiments identical in spirit to them have actually been performed, and not just with electrons but also with light, and not just once, but over and over again, by scientists in laboratories throughout the world—these experiments have forced upon us a new view of the nature of existence. They have created a new science, a science just as rigorous, just as exact as any other, but far more profound. What shall we call this new discipline? The philosopher and physicist Abner Shimony has named it aptly: It is the science of experimental metaphysics.

Buried within us all is the unalterable conviction that the physical world exists as something objectively real, and independent of us. Everyone is familiar with the old chestnut about the tree that falls in the forest with nobody present. The philosopher wants to know if it made a sound; the rest of us have a tendency to snort in derision at such a question. Certainly I do. As a physical scientist I have no truck with such stuff. Or at least I used to feel that way, before commencing that series of experiments. But now . . . now I am not so sure.

Tables, chairs—they exist apart from our perception of them, and they maintain their identities through thick and thin. They do not rise up and vanish in a puff of smoke every time I leave the room, only to return as I stroll back in again. This quality of permanence is the identifying characteristic of anything I am prepared to label as being objectively real, as opposed to a hallucination. But in the world that the new science of experimental metaphysics has revealed to us, this permanence has entirely melted away. In that world nothing is objectively real independently of us. Rather, *it is the observation itself that brings the physical world into existence.*

The University of Texas physicist John Wheeler, one of the leaders of the effort to comprehend the subtleties of quantum theory, has coined the term *participatory universe* to refer to this strange interconnection. The act of observation creates the electron. Prior to its detection in the cloud chamber this particle simply did not exist. On the other hand, once detected it persisted as a real member of the physical world. This is why that electron left its track in not one but two cloud chambers, the one behind the other, so giving evidence of its trajectory through space. Furthermore, whenever a chamber is placed in an electron's path a track will form— is guaranteed to form, whether anyone wishes it to or not. To this extent the electron possesses the permanence of the objectively real. It is no hallucination.

At the very same time, though, the act of another kind of observation—or participation—creates the wave. Apart from observation, therefore, the electron has no objective reality at all. It has merely what might be called a set of potentialities, any one of which can be called into being. And as for the delayed choice experiment, it does not really alter the past at all. Rather, the past did not even exist until brought into being; brought into being in the flicker of an eyelash, in a sacred act of creation—a small creation to be sure, even a paltry one, but creation all the same: the sudden appearance of a track in a cloud chamber in a musty laboratory on New Year's Eve.

14

Schrödinger's Cat

The argument of this and the preceding chapter is that in the fitness for life of the cosmos we are witnessing the effects of a gigantic symbiosis, a symbiosis between the universe on the one hand and life on the other. The proposal is that the cosmos brought forth life in order to exist. The first half of the argument is now complete: It is that in order for a single particle to exist, it must be observed. In the present chapter two additional steps will be taken. The first is to argue that what is true of a single particle is also true of collections of particles: stones, planets—even the universe as a whole. The implication is that the very cosmos does not exist unless observed. And the second step will be that *only a conscious mind is capable of performing such an observation.*

Creation takes minds. But it is only living beings that have minds. That is why a symbiosis is at work. That is why the cosmos on the one hand and life on the other are melded into a single unity. Living beings cannot exist without an appropriately fit environment. But that environment cannot exist without living beings to observe it.

What is an observation? The question looks simple. An observation is the appearance of a line of droplets in a cloud chamber. An observation is the exposure of a strip of photographic film. Surely these are not so difficult to analyze. Surely no intervention from a conscious mind is necessary here. But if there is any moral to the preceding series of experiments, it is that nothing is as simple as it seems. If the electron can be both a particle and a wave, if it can both arrive and not arrive somewhere, then why could not an observation both be made and not be made? If the microscopic world exhibits so maddening an ambiguity, then why not the macroscopic world of measurement as well?

One of the problems is that English is not a good language for the discussion of these matters. The experiments recounted above appear to involve us in logical paradoxes. A close look shows that they are not devoid of logic, though. Rather, it is our very language that is at fault. English is admirably suited to the description of the large-scale world of daily experience, but it fails utterly when faced with microscopic phenomena. The statement, for example,

> The electron is a particle.

is shorthand for:

> The electron is, and it is a particle.

—an assertion whose initial part is open to doubt. It is here, in the structure of our language and its insistence on imparting objective reality to the world, that much of the seeming irrationality of the electron's behavior arises. Our difficulties are primarily linguistic in nature. To escape them, what we need is a new language.

The name of that language is quantum mechanics.

Quantum mechanics is the theory of the submicroscopic

world, and only in its terms can a proper understanding of that world and its workings be attained. It was developed in an immense effort extending over the first three decades of this century, culminating in the introduction of the famous wave equation in 1926 by the Austrian physicist Erwin Schrödinger. Schrödinger had seized upon a suggestion put forward two years earlier by the French physicist Louis de Broglie that to every particle there was associated a wave, and asked himself what mathematical expression would govern the propagation of this wave. The Schrödinger equation was his answer: It is the fundamental law of quantum physics.

Schrödinger gave his wave the Greek symbol ψ, pronounced *psi*. In many ways it is an ordinary wave. It has crests and troughs, and exhibits the phenomenon of interference. The difference comes in the interpretation. What does ψ represent? In his initial thinking Schrödinger had inclined toward the view that it described some sort of actual wavelike structure of the particle, but subsequent research showed this interpretation to be untenable. Rather, it measures only the *probability* of the particle's being present. More precisely, *at every point in space, ψ squared is the probability that the particle is located there.*

For example, the electron's psi function in the two-slit experiments described in the previous chapter undergoes constructive and destructive interference, and so exhibits the characteristic interference pattern sketched in Figure 64 (page 206). The graph along the right-hand edge of that figure illustrates the wave function's amplitude along the line of detectors. What does it say about the results of the experiment? The electron is most likely to be found where the probability is greatest; this is at the midpoint directly across from the two slits. As one progresses up or down from this point, the wave function quickly drops to zero, implying no electrons at all at these locations, then increases, implying a reasonable probability of finding electrons, then drops to zero again, and so forth. If a large number of electrons are sent through the slits, they will preferentially accumulate in the regions where ψ is greatest, totally avoiding those where it is zero. The strip of photographic film will then be exposed accordingly, thus yielding the telltale interference pattern. In this way the predictions of the theory are found to be in agreement with the results of the experiment.

Note, however, that the predictions are only probabilistic in nature. Quantum mechanics never guarantees anything with certainty. It never says where the electron is assured to be. The interference pattern of Figure 64 is therefore not the only possible outcome of the experiment. It is just the most likely one. Flipping a coin, one has equal chances of getting either heads or tails; a large number of flips most likely will yield equal numbers of each. On the other hand, it is always possible to run into a string of lucky breaks, and finish with nothing but heads. Similarly, there is always some chance of getting an unexpected result in the interference experiment. One can insert a strip of photographic film and expose it, then insert a second and expose it as well. One can repeat the identical process a hundred times and obtain the same result—but then, on the hundred and first, find something totally different. Only the right-hand end of the film might be exposed, for instance, or only the left. The pattern obtained might turn out to bear no resemblance at all to the ordinary interference pattern. In terms of the actual numbers of electrons involved in an experiment, the probability of this occurring is exceedingly low, far lower than that of flipping heads a hundred times, and for all practical purposes the possibility can be neglected. But the matter is one of principle, and the principle is that nothing can be predicted with certainty.

The implication is that the very concepts of cause and effect have vanished. If an anomalous result were to be obtained in one out of an identical sequence of experiments, one would naturally seek to explain the discrepancy. One would ask what made for the difference. But according to quantum mechanics, nothing made for the difference. Indeed, the very idea that a cause might be found is contrary to the spirit of the theory, for to say that a cause exists is to imply the experiment ought to have a certain outcome. But there are no "oughts" in quantum mechanics. There are only probabilities. The theory resolutely refuses to make any statements of an absolute nature.

In this regard quantum mechanics is utterly unlike the classical mechanics that governs the behavior of things in everyday life. If I were to witness a string of one hundred heads flipped by the croupier of a Las Vegas casino, I would not shrug my shoulders philosophically—I would call the police. I would be convinced the coin was loaded. This conviction arises because in the large-scale world of daily experience things have causes, and unforeseen events can be accounted for. The idea makes sense. But it must be abandoned in the ambiguous world of the quantum.

The implications are so extreme that many physicists found themselves unable to accept quantum mechanics in the early days of its creation. Ironically enough, the most prominent of the dissidents was a man not known for his adherence to outmoded forms of thought: Albert Einstein. Indeed, Einstein played an important part in the very creation of quantum theory. Nevertheless he later rejected it, and he responded to Schrödinger's probabilistic wave function with the famous aphorism "God does not play dice with the universe."

But he was wrong. Those dice are utterly essential to a proper understanding of subatomic behavior. They are the means whereby quantum mechanics is able to encompass the endlessly slippery quality of the microscopic world. Any theory that made absolute predictions could only do so by attributing a hard, objective reality to this world, in conformity with our classical concepts of cause and effect. It is precisely because of its steadfast refusal to speak in such terms that quantum theory does justice to the ambiguities of existence revealed by experimental metaphysics.

As an example, consider the experiment recounted in the previous chapter designed to test the "traffic jam" theory of the interference pattern. In that experiment the electron gun was adjusted to emit electrons so slowly that each traversed the apparatus alone, thus eliminating the jam. Nevertheless, the ordinary interference pattern was obtained. From the normal point of view, such a result is incomprehensible. In terms of quantum theory, on the other hand, the mystery disappears; for no matter how strong or weak it may be, a wave is a wave, and so naturally exhibits interference. But note the complete disappearance of objective reality here. How are we to visualize the electron's path in this experiment? In the mind's eye an image arises of a particle—some kind of tiny golf ball—leaving the gun, traversing the slits, and winding up somewhere along the bank of detectors. We cannot help thinking in such terms, for we are bound by our nature, by the language we speak, to consider the electron a thing. The experiment makes abundantly clear, however, that no such path ever existed. And the theory works in full conformity, by refusing to describe the electron's trajectory through space. Nowhere does quantum mechanics explain the means whereby this indivisible unit has managed to make its way through two slits at once.

The next experiment, though, did succeed in detecting the electrons as they traversed the apparatus—by mounting additional cloud chambers behind the slits. Several striking results were obtained. In the first case, the electrons now acted the way particles ought to act, each traversing one and only one slit and winding up in the appropriate element of the detector bank, and so exhibiting a clear sense of a path through space. On the other hand, the telltale interference pattern was replaced by a different and quite unrelated one. All in all, the act of observing the electrons seems to have decisively altered their behavior.

As indeed it did. *The act of observation gave reality to the particles.* There is no contradiction between these two experiments. Rather, the quantum world remains ambiguous only so long as it is not detected. Once it's detected, however, every trace of ambiguity is removed from its behavior.

What happens to the wave function ψ in this process of detection? Figure 72 diagrams its behavior. At the instant illustrated in this diagram the upper of the two cloud chambers has just sprouted a track, revealing which slit the electron has traversed. *At this moment the wave function changes.* Why does it change? Because the psi wave measures a probability. Before detecting the electron there was no knowledge as to where it was, and so ψ extended everywhere. But by detecting the electron we have learned exactly where it is: It is by the upper slit, and it is not by the lower slit, and the wave function must now reflect this fact. Suddenly there must be a new wave function, exhibiting a probability of zero that the electron can be anywhere other than where it is known to be. For example, ψ can no longer extend down to the lower slit; for if it did, a subsequent measurement made the barest fraction of a second later would have a reasonable chance of finding the electron there. Indeed, ψ can no longer extend anywhere beyond the location of the upper cloud chamber. It must have been transformed into a single hump, concentrated at the point of detection, as illustrated in Figure 72.

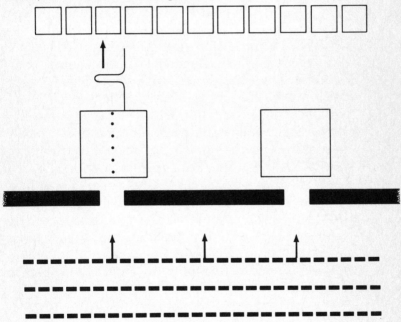

Figure 72

This sudden change in ψ, induced by the act of observation, is referred to as the collapse of the wave function. Once it has collapsed in this manner, ψ retains its humplike form, and it progresses in a straight line from the slit to the detector bank, thus imparting a well-defined path to the electron's motion. So the electron comes into being through its detection.

*   *   *

Return now to that detection. Why do I claim that a conscious mind is required to bring it about? Why could an inanimate measuring instrument not do just as well?

Analyze the means whereby it takes place. An electron falling upon a cloud chamber ionizes an atom, which then acts as the impurity about which a droplet of water forms. The incoming electron's wave function now collapses into a hump, so imparting a well-defined path to the particle's motion; it proceeds on to the next atom along that path, ionizes it, and so the process continues. As for the strip of photographic film, its operation is in essence no different, for here too the exposure is caused by the ionization, or at any rate the excitation, of an individual atom by the incoming electron.

That's a good description . . . but it's in the wrong language. It is in *English*. Translate it now into quantum mechanics. As for the electron, it is described by a Schrödinger wave, ψ. But how does the theory deal with the second component, the atom whose ionization triggers the detection?

The theory assigns a separate wave function to it. A second name is required, to distinguish this new wave from that of the electron. Call the atom's psi function A. A has wavelike properties, analogous to ψ's, describing where the atom is likely to be found. But unlike ψ it also has two further parts, expressing the fact that the atom can be found in either of two different states: the ordinary intact state or the ionized state. Give these parts the symbols A (intact) and A (ionized).

The full wave function describing the atom is the sum of these two:

$$A \text{ (intact)} + A \text{ (ionized)}$$

and, as before, A (intact) squared is the probability of finding the atom in its ordinary intact state. Similarly, A (ionized) squared is the probability that the atom is ionized—that a detection has been made. Quantum theory describes how to calculate these two parts of the atom's wave function. In such and such an experiment, for example, it may turn out that the atom has a 70 percent chance of being ionized by the electron. But that is *all* the theory predicts. It never predicts when the atom ionizes, nor which among all those filling the cloud chamber will be the one to do so. Just as a gambler might someday flip a long string of heads, so the atom is ultimately free to do anything, 70 percent probability or no; and just as quantum theory refused to describe the passage of the electron through the two slits, so now it refuses to speak more specifically about the process of ionization. As, indeed, it must—for the atom too is a quantum system, and is subject to the same degree of slipperiness as the electron. Just as an electron can both pass and not pass through a slit, so an atom can be both ionized and intact.

But wait a minute here! Where are all these extra wave functions coming from? There is no particular interest in that atom. The atom is secondary. Its only function has been to bring the electron into existence by detecting it. But suddenly the atom itself seems to be evaporating away.

Nothing for it—the atom too must be brought into existence. It is necessary to observe the state of the atom, find out whether it is ionized or intact. The instant this is done its wave function will collapse—not into a hump, but into either A (ionized) or A (intact). Only then will the electron finally be nailed down. And how to perform this observation? Look for that telltale droplet of water condensed about the ion.

Quantum mechanics describes the process of condensation. Of course we must use its language. We must invent yet a third psi function, W, say, representing the state of the water in the cloud chamber; and it too will have two parts, W (vapor) representing the water uniformly spread about as humidity, and W (droplet) representing the droplet condensed about the ion. The full wave will be the sum of these two. Months later, exhausted, bleary-eyed from pages of mathematics, we have the answer in hand: a probability of 95 percent, no more and no less, that a droplet will collect about an ion.

That's not enough, of course. What's needed is a precise answer, something we can hang on to, an escape from this endless quantum ambiguity. After all, what about the remaining 5 percent probability of not getting a droplet? How to be sure the droplet formed? We can't. We can't unless we detect it. But . . .

An infinite regress is in the making.

*   *   *

SCHRÖDINGER'S CAT

To detect the droplet, look at it—look with eyes, using light. That light, of course, is itself a quantum entity, subject to the same degree of ambiguity as the electron. Generated by a light bulb, it flies off, reflects from the water droplet, and so enters the experimenter's eyes, carrying with it information concerning the presence of a track. It strikes the experimenter's retina, and there induces an alteration in the molecules responsible for visual perception, an alteration fundamentally quantum-mechanical in nature. The net result is a nerve impulse propagating down a physicist's optic nerve, the propagation a complex matter with both electrical and chemical aspects, but itself subject to quantum laws. The impulse enters a brain, and there in some unknown manner is translated into perception. But this too cannot violate any principles of physics. So where does the regress terminate? At what point in the complex chain of interlocking wave functions does the detection occur?

No one knows.

No one knows because the language of quantum mechanics possesses no means of describing the final, absolute registration of an observation. The theory speaks the language of wave functions and probabilities, and it speaks this language only, rigorously avoiding any other form of discourse. But in order to speak of an observation, one must find some means of escaping from these ambiguities. One must pass beyond quantum mechanics.

The theory's inability to deal with this question is best appreciated with the help of an example first proposed by Erwin Schrödinger. Schrödinger's is not an experiment to be performed. Nor does it need to be performed. It is a hypothetical experiment, a mental exercise designed to illustrate the extraordinary nature of quantum theory most clearly.

The experiment consists of a box on a table, two feet on a side and painted a funereal black. Its door is bolted firmly shut. Projecting from the box is a length of pipe leading into a nearby closet, the pipe blocked in its middle by a shutter. Something might be in that closet; if so, when I open the shutter it is going to flow down the pipe and into the box.

A stopwatch is in my hand. As I set it ticking, I reach out

and open the shutter. As the single minute allotted to the experiment ticks away, I pace about nervously. Finally, at the appropriate moment I replace the shutter. I unbolt the door. Swinging it open I bend down and peer inside.

In that instant, something happens.

Inside the box is a cat. Inside the closet are (maybe) electrons—but only a few of them, at most two or three wandering about. When the shutter is opened one of those electrons might (or might not) find its way down the pipe and into the box. And there, inside the box, waits the most devilish of devices: an instrument that, when triggered by an electron, spews forth a deadly dose of poison gas. Open the shutter for one minute and then close it again. Is the cat dead or alive?

Quantum theory describes the cat by a wave function, C, say. It has two components, one representing the living animal and one the dead. After the experiment is completed, the wave function is:

$$C \text{ (alive)} + C \text{ (dead)}$$

What is the meaning of this wave function? It is that the cat has such and such a probability of being alive, and such and such a probability of being dead. And that is all the theory has to say. It never tells us whether the animal lived or died, and it never will.

What to do about this limitation of quantum theory? How to surmount it? Several ideas have been proposed.

The first idea is the one held by most physicists. It is based on the recognition that in every observation there occurs at some point a shift in scale. There is a transition from the world of the exceedingly small to the far larger world of everyday things. In the case of the cloud chamber this transition occurs at the moment an ion, a microscopic object, triggers the formation of a water droplet, a macroscopic object big enough to be seen with the naked eye. In the case of Schrödinger's cat it occurs within that devilish device which, upon receiving a single electron, releases an enormous number of poison atoms into the air. *And the nice thing about the large-scale world is that you can depend on it.*

In the everyday world all the old notions of cause and

effect, of the existence of objective reality aside from our intervention, and of the permanence of things—all hold true. Every time someone climbs out of bed these principles are tested and found adequate. People's arms never fall off of their own accord as they walk about the room. Airplanes' wings do not suddenly cease lifting the plane. Surfers on the waves of Waikiki never find themselves riding the crest of some mysterious collection of particles. In these and countless other ways we rely throughout every minute of our lives on the orderly functioning of the world, including the natural world and the world of machines that we have constructed; and every minute of our lives this confidence is justified by the world's evident conformation to our expectations. The large-scale world obeys laws.

Clearly the essential lawlessness of quantum mechanics does not penetrate into this world. Thus the theory must have no relevance to it. The conclusion is that *an observation occurs, and reality is created, in the instant at which that transition in scale occurs.*

An electron may not exist in the ordinary sense of the word, but a cat is either dead or alive, and it remains so whether someone is watching or not. If a cloud chamber develops a track, the presence of an observer is irrelevant to that fact. Thus the electron wave is detected by the exposure of a strip of photographic film—even before that film is examined, and even if the film is tossed into a wastebasket without ever being developed at all. A nuclear war might well wipe out every trace of life on Earth; but if by chance some cloud chamber manages to survive the holocaust, it will still go on recording tracks unobserved, each time bringing an electron into existence.

That is the view held by most physicists of the nature of observation in quantum mechanics. Two points stand out. The first is that according to this view the universe as a whole is not subject to the maddening ambiguity one finds in the microscopic realm. The second is that inanimate machines are perfectly capable of performing observations. Thus, according to the most widely accepted picture, observation is not required to bring the universe into being. Furthermore, even if it were, consciousness would not be required to do so.

If this standard picture were correct, the argument for a symbiotic universe would evaporate into thin air. But it may not be correct. There are many difficulties with it.

In the first place, it is an error to claim that quantum mechanics is exclusively concerned with submicroscopic phenomena. In truth it extends into the large-scale world as well. The existence of electrical conductors such as the copper wiring in one's house is an example. Another is the laser, yet another the electronic chips of which modern computers are built. These things, all immense compared to an atom, are nevertheless quantum-mechanical in their nature, and they cannot be understood on the basis of classical concepts at all.

Furthermore, quantum mechanics blithely refuses to admit its insufficiency in the large-scale world. The theory claims to be valid for all things, large and small alike. Nowhere in its principles is found a statement to the effect that it only deals with small things. According to it, everything is ultimately quantum-mechanical in nature. Electrons are not the only things that are both particles and waves. People are too.

But if we possess psi waves, where are they? Why doesn't our quantum nature reveal itself? The answer is twofold. On the one hand, the length of the Schrödinger wave associated with anything as large as a person is extraordinarily small, too small to make its presence known. On the other, an individual particle may act randomly but large numbers of them behave according to strict laws—just as each individual worker in a large city enjoys free will, but rush hour comes at 5 P.M. all the same. The abandonment of cause and effect is masked by the very great number of particles (atoms, electrons) making up the world of everyday experience.

If things were otherwise we would experience ourselves as what at heart we are: quantum structures. But while the theory penetrates into the large-scale world, the absolute, irreconcilable difference remains between its view of the nature of reality and the set of conceptions we carry in our heads. One must choose between the two. Those physicists who hold to the idea that machines can perform observations have opted for the traditional view of reality. But they are then faced with the necessity of explaining why the insight into the nature of existence provided by the science of experimental metaphysics does not carry over into the everyday world. No convincing answer to this question has ever been given. Furthermore, these physicists must also decide just where in the range of sizes the transition between the small and the large occurs. If a droplet of water in a cloud chamber is objectively real, then how about something 10 percent its size? Does that exist if no one is there to see? If an observation is required to bring a single electron into being, how about five electrons? Five hundred? No one has an answer to these questions.

In recent years an important insight has been achieved with the help of a device cheerfully known as the SQUID— the Superconducting Quantum Interference Detector. The SQUID has enabled us to perform, on objects big enough to be seen with the naked eye, experiments that previously had been confined to the submicroscopic realm. This device has explicitly demonstrated that large-scale objects have both particlelike and wavelike properties, and that they possess the same slippery, ambiguous nature as the electron.

These experiments have made clear that the transition between the quantum and classical worlds, whatever its nature may be, has nothing to do with size. The process of observation may indeed involve a change in scale, but the fact is irrelevant. Just like the particle and the wave, nothing else exists until it is observed. Schrödinger's cat waits before me in its box. I want to know whether that cat is dead or alive . . . and I am convinced that the cat *is either* dead or alive, and I am frustrated by the refusal of quantum mechanics to tell me which. I regard this as a problem with the theory. But the theory has no problems. It is I who have a problem: a problem in assimilating the tremendous significance, the profound meaning of the discoveries of experimental metaphysics. I have been all too willing to accept the successes of quantum theory, the advances in knowledge it has provided, while at the same time refusing to accept its message. And the message of the quantum is that the cat is not either dead or alive, but that it is *neither dead nor alive*. The state of health of the animal has no existence until brought into being by an observation.

And how is that observation performed? Eugene Wigner, physicist at Princeton University and recipient of the Nobel Prize for his work on quantum mechanics, has advanced the following remarkable prescription: I look.

I look. And I see. In that act of seeing, something enters my mind: not light impinging upon a retina, not nerve impulses flowing into a brain, but *knowledge*. And the remarkable thing about knowledge is that it is not subject to the ambiguities of quantum mechanics. If I become conscious that a cloud chamber has developed a track, I do not at the same moment become aware of any uncertainty concerning the fact. There is no perception of the track's simultaneous existence and absence. If I am aware of sitting down I am not simultaneously aware of standing up. It is only here, Wigner argues, that the regress of quantum theory terminates and certainty appears: in that final step in which the nerve impulses of a brain are translated into the awareness of a mind. Consciousness is all we have left with which to create the universe.

It's a participatory universe; nothing exists unless it is observed. But observation itself is problematic. According to quantum theory, the same slipperiness, the same endless, maddening ambiguity that surrounds the electron also surrounds the act of observation that brings it into being. The theory denies the very possibility of making one.

And yet we know, with absolute, irrefutable certainty, that observations take place. We know this by the evidence of our senses. Every moment of our waking lives we perform a multitude of them: You see the page upon which these words are printed, I hear the passing truck. But an experience such as seeing or hearing is a mental event. While initiated by physiological events in a retina or an eardrum, the ultimate awareness takes place within a mind. Wigner's argument is that *mind therefore cannot be bound by the limitations of quantum mechanics*. The electron changes from potentiality to actuality only when I become aware of a line of droplets in a cloud chamber. The past one minute of the history of Schrödinger's cat does not spring into being until I peer inside its box.

The very cosmos itself depends for its being on the uttermost mystery of consciousness. And thus the symbiosis, the union between the physical world and mind, the great metaphysical dance by which each brings into being the other.

The box on the table before me is two feet on a side and painted a funereal black. Its door is bolted firmly shut; someone has scrawled upon it the symbol ψ, first letter of the Greek word *psyche*. Projecting from the box is a length of pipe leading into a nearby closet, the pipe blocked in its middle by a shutter. I am gripped by a certain tension. A stopwatch is in my hand, and as the single minute allotted to the experiment clicks away, I pace the room nervously. Finally, at the appropriate moment I replace the shutter. Bending down, I unbolt the door and peer inside.

In that instant something happens: The aliveness of the cat springs into being. It walks out, purring and blinking its eyes.

Epilogue

Ocean Voyage

There is not a breath of wind. There is not the slightest sound. That chipmunk in the bushes seems to have closed up operations for the night. Overhead, the stars are strewn across a darkness, a blackness so profound that for a moment, for the barest flicker of an instant, I can almost sense their inconceivable distance. In a sudden, exalting burst of vertigo I fancy what it would be like to fly, to fall up and into that ocean. And in my imagination I am falling now, falling slowly, falling endlessly, tumbling gently through the stars in the great and perfect isolation of the night.

My goal in this book has been to sketch the outlines of my state of knowledge. The person the reader finds standing there in his backyard, lost in exhilaration as he gazes upward into that fathomless immensity of sky, is a person who knows certain things. In the last fourteen chapters I have summarized those things. Two great discoveries have been recounted: first, the fitness of the cosmic environment for life, a fitness that hangs by a thread and that cries out for explanation; and second, quantum mechanics, with its extraordinary insight into the role played by observation in the creation of reality. Taken together, these two point to yet a third: to the notion of a symbiotic universe. In some strange fashion, those stars up there are not autonomous entities. Reason enough for exhilaration.

But science is the art of doubt. I have already emphasized that research, all research, is hedged about with uncertainty. And at the frontiers of knowledge, at the very limits of our understanding, often we encounter little more than massive clouds of confusion. And why should this not be so? Surely no great discovery ever is won with ease. Surely no profound revolution in thought yields itself up without a struggle.

Chapter 12 recounted some of the uncertainties surrounding the concept of a universe fitted for life. Many of the arguments that have been advanced in support of this concept are subject to doubt. In that chapter I commented that none of those uncertainties struck me as being particularly serious. Rather, they were the sorts of hurdles that people surmount regularly. Unfortunately, the difficulties mentioned in that chapter were but the tip of an iceberg. I have kept the toughest stuff for last. This person standing here in his backyard, sniffing lilacs and apple blossoms, is in deep trouble right now. He is lost in clouds and mists, the uttermost vapors of confusion. He is faced with one problem that for more than half a century has resisted the efforts of the world's best physicists, a second whose solution will entail a program of research sure to present the most hideous difficulties. And more than that, he also must deal with questions lying at the core of what we mean by science—with the very nature of physical reality itself.

Quantum mechanics is one of the glories of our age. Since its discovery the theory has proved spectacularly successful, moving from triumph to triumph in an unbroken line. Its influence is decisive throughout the physical sciences. And with every new success, every fresh confirmation by ever more refined experiments, we gain yet further conviction that the theory is right.

And yet surrounding quantum mechanics there lie the most profound enigmas. So subtle are the questions raised by the theory, so obscure its foundations, that even today, more than half a century after its formulation, there is grave uncertainty concerning its interpretation. What does quantum theory mean? What is it telling us about the nature of reality? These questions have been debated for generations, and we seem no closer to a resolution now than ever.

I have already described the severe difficulties facing the majority view, which holds that observations do not require minds. Now is the time openly to acknowledge the equally severe difficulties facing the minority view, that they do. Wigner's argument rests on the assumption that mind is not subject to the limitations and ambiguities of quantum theory. Perhaps it is not. But surely this is close to arguing that mind is not subject to any form of physical causation at all.

We have, of course, abundant evidence to the contrary. So simple a matter as missing a meal is a good example: The absence of food, a physical state, is accompanied by mental sensations of hunger, irritability, and the like. A stiff shot of the physical substance known as whiskey, on the other hand, reverses the situation. Psychologists have found that electrical stimulation of the brain can dramatically influence the function of the mind. According to one researcher, a man who had just attempted suicide reported, upon being stimulated, that he "suddenly felt good"; other subjects have reported intense sexual feelings. Sodium pentothal, the so-called truth serum, achieves its effects by reducing the subject's sense of anxiety. Its mechanism of action is well understood: The drug acts by enhancing the effects on one nerve cell of an impulse arriving from another. The more such examples accumulate, the less and less reason we have to believe in any fundamental dichotomy between mind and matter.

Wigner's "mentalist" interpretation of quantum mechanics comes close to holding that mind stands outside all the rest of nature. It is difficult to believe that this could be the case. On the one hand, such a view smacks of the super-natural, and it goes against the fundamental presupposition that all things behave according to the dictates of natural law. On the other hand it goes against the lesson of centuries of history: Time after time, claims have been made that such and such a phenomenon violated some law of nature and could not possibly be accounted for on the basis of ordinary principles, and time after time these claims have proved unfounded.

Where is one to draw the line between physical things, to which quantum theory applies, and minds, to which it does not? The more we have learned, the more it has become evident that no such line exists. It seems more likely that what we call mental activity lies merely at one end of a continuous spectrum. Surely mentally retarded people have minds. Surely animals have rudimentary minds as well. A cat stalking a bird is aware of its prey. It could be conditioned to respond to the sudden appearance of a line of droplets in a cloud chamber. Cannot, then, a cat bring electrons into being by its observation? And if so, why not an amoeba or a bacterium?

All in all, it seems foolish to single out one end of a spectrum as fundamentally different from the other. Recent advances in artificial intelligence have sharpened this point. Computers can be programmed to recognize spelling errors and perform medical diagnoses, guide unmanned spacecraft to the deepest reaches of the solar system and play daunting games of chess. Is this evidence that computers think and have minds? Many of us might be unwilling to take that step, but are we perhaps merely prejudiced in this regard? Why couldn't a machine, a thing of wires, transistors and electronic chips, exhibit thought? Put it this way: Computers can do things that we do only by thinking about them.

The "mentalist" interpretation of quantum theory runs into severe difficulties in dealing with situations involving more than one observer, difficulties that do not plague the majority interpretation. Imagine two different people watching the box containing Schrödinger's cat—you and me. You crouch down directly in front. I, on the other hand, will stand off to one side, out of view of the door. Now open the box. As for you, you see the cat right away. But as for me, all I can see is your face. All I see is the sudden relaxation of your brow, the slow appearance of a smile upon your lips. When, then, does the aliveness of the cat spring into being? When you see it? When I guess from the expression on your face? Or only when you reach forward, pick up the animal, and hold it aloft for my inspection? No one has an answer to these questions.

Throughout this book, I have repeatedly compared the condition of life in the cosmos to that of a man before a firing squad composed of defective rifles. But here too a vagueness and an indistinctness has crept in. Perhaps the reader will have noticed it.

The problem is that, strictly speaking, there is no such thing as a defective firearm. Each gun does only what it must. After all, none of them have any choice in the matter. All are bound by the nature of their construction: The cocking mechanism of one has developed a certain amount of play; the firing pin of another is jammed by dust. While we might be tempted to call the misfiring of a great many rifles a happy accident, it is not really so at all.

All the "coincidences" I have pointed to, and to which I attach so much importance, are in fact not coincidences at all. They are necessary consequences of the laws of physics. It is within these laws that the fitness of the cosmos for life arises. It is to them, and not their consequences, that we must look. I have persisted in claiming these laws to be uniquely suited to life. Suited they are—but who is to say they are uniquely suited? Who is to say that some completely *different* set of natural laws would not also have led to life?

Were the laws of nuclear physics different, Hoyle's resonances between helium, beryllium, and carbon within the red giants would not occur. But recall that a tuning fork possesses not one but many resonances—its overtone series. So do nuclei. If these laws differed, could perhaps some other saving resonance come into play? Or again, Whitrow argued that had space more than three dimensions, no planet could orbit its sun stably. His argument relied on the fact that gravitational attraction can be thought of as arising from lines of force. But other attractions known to physics cannot be thought of in these terms. What if gravitation acted like one of these? Might not planets stably orbit their suns in four dimensions in such an eventuality? Or yet again, the fortuitous matching between the color of the Sun and the absorption properties of chlorophyll rests upon the principles governing their respective structures. If, for example, the attraction of gravitation were stronger, that matching would be spoiled . . . or can we simultaneously imagine some counterbalancing shift in the strength of the attraction holding together chlorophyll?

How to answer such questions? There is only one way: to take those other laws of nature seriously, and to analyze their consequences in detail. We must imagine an alternate reality, a universe complete within itself but differing from our own in some particular physical principle, and we must ask whether life could have arisen in such an eventuality. Then we must go on and create a second alternate reality, differing in another physical law, and then yet a third, and so on, each time evaluating the suitability of our hypothetical cosmos to life. We must build in our minds a universe of universes.

Such a program of research has never been attempted. There's good reason: It is going to be extraordinarily difficult. The whole task of science is to comprehend but a single universe, the real one, and even this has proved hard enough for anyone. We're nowhere near completion on the project. To comprehend some other universe would be equally hard— and in such a task we would no longer have the benefit of reality as it is actually given to us against which to measure our speculations. We would be like men blind from birth trying to imagine colors. Furthermore, it is required to comprehend not one, but myriads of alternate realities in this way.

No—it will be no easy task. But we have no choice. It is the only way forward. The widespread indifference most scientists feel to the concept of a cosmos suited to life springs from many sources, but one of them I have not yet mentioned. It is that the concept has never been expressed in terms of numbers. So far it has proved impossible to quantify the fitness of the environment, to measure it precisely. How do you assign a number to anything as ill defined as habitability? But numbers have a hardness and a sharpness about them that carry great force. They are the indispensable tools of science. Lacking these tools, arguments for or against the fitness of the environment will forever remain vague, unsatisfying, and indistinct.

Only this program can provide us with those numbers. It will tell us how fit is fit. Does the actual cosmos rank in the most fit 0.001 percent of all conceivable possibilities? That would lie far beyond the limits of coincidence. It would constitute clear proof, persuasive to all, that some explanation for the habitability of the cosmos must be found. The same would be true if actuality ranked in the top 1 percent, or 5 percent. At the same time, of course, the opposite result is also possible. A full 90 percent of all cases, when examined with sufficient care, might exhibit enough of those wild coincidences to yield life. Though I myself believe such a result to be unlikely, surely it is not inconceivable.

And in such a case, the reader would be well advised to toss this book unceremoniously into a wastebasket. Nothing would be left to explain. There would be no mystery, no string of astonishing coincidences, no need to invoke some weird and speculative symbiosis between mind and cosmos. The fitness of the environment would be reduced to a simple matter of necessity; and what might have been a great revolution in thought would dissolve into a chimera, a trick of shifting clouds.

It is human nature to see patterns where none exist. Our minds seem to be constructed that way. Those constellations above my head right now are a good example: I persist in seeing shapes within them, shapes that have no counterpart in reality. One reminds me of a dipper, another of a swan.

Chapter 11 mentioned that the geometry of the cosmos might not be that of Euclid, but rather that of figures drawn on a sphere. A striking feature of such a geometry is that it invites us to think of *two different spheres*—or 2 million.

Within this analogy each globe would correspond to a universe, each one complete within itself but separated from every other. We, crawling like ants over the surface of our sphere, would have no way of reaching any of the others. No spaceship could voyage away from this cosmos and toward a second. No matter how far we peered with telescopes we would be incapable of seeing these alternate universes. No measurement performed in the laboratory could reveal the slightest hint of their existence. The analogy of the spheres is not a good one in this regard, for it invites us to imagine hopping from globe to globe. But in the analogy, the spaces separating the various globes do not exist. There simply is no path leading from one to another—they form utterly disconnected geometries. Such disconnected universes can also be envisaged within the context of ordinary Euclidian geometry. The model for that geometry is a flat piece of paper; imagine not one but an entire sheaf of them.

Is it conceivable that such universes could operate according to laws of physics different from our own? Perhaps it is. Perhaps there exist hundreds, millions, even an infinite number of them, each unrelated to the others and possessing its own unique set of laws. In one such cosmos the electron charge would not balance that of the proton. In another it would, but each would be just twice as big. In yet a third, water would make an even better solvent. Some of those universes would be sterile. But others would be capable of supporting life.

All this bears a strong resemblance to the notion discussed above of alternate sets of laws. The difference is that these sets are *real*. Even though we have no way of making contact with them, the other universes are considered to exist. The earlier discussion, on the other hand, dealt entirely in terms of a single reality. It was of a what-if nature, asking questions similar to "What if Napoleon had never lived?" And this difference has enormous significance.

Chapter 5 went to great lengths to argue that it would be wrong to expound upon the habitability of the Earth—the fitness of our planet's temperature and the great amounts of water here. To do so would be akin to a flower's expostulating on the virtues of its niche. In truth the Earth is merely a somewhat larger niche, and were it not fit for life some other planet would be and could well be inhabited. In contrast, the fitness of the universe is another matter altogether, for if the laws of physics did not conform to life's requirements, no location would be habitable. Life would be absent—not just here or there, but in reality.

But were the cosmos not single but multiple, and were different universes to possess different laws, then what we know of these laws would not refer to all of reality. It would merely refer to this particular set of circumstances—to our niche. The similarity is obvious between a set of separate geometries, each constituting its own reality, and a collection of planets, each either fit or unfit for life. If one reality is not a good niche, life simply flourishes in another.

My program of research outlined above could equally well be applied in such a multiple cosmos. The actual work, in fact, would be identical. The only difference would lie in the interpretation. The program would tell us how many of those other universes were fit abodes of life. But the significance of that information is not very great. Who cares if a mere 0.001 percent out of the list of all possible universes turn out to be habitable? This one is . . . and one is all we need.

If reality is multiple, the entire argument of this book tumbles yet again into oblivion. Once more, nothing would be left to explain. What all of us believe to be physical reality would merely turn out to be our local set of conditions, and the fitness of these conditions for life would be no more surprising than the presence of some soil in a cleft between two boulders. And what I thought was a profound discovery would again dissolve into a trick of shifting clouds.

There is a second way in which the concept of multiple realities could come into play. It involves the possibility that the present expansion of the universe is ultimately destined to reverse into a contraction. No one knows what came before the Big Bang. Similarly, no one knows the fate of a contracting universe once it has collapsed all the way down to a second Big Bang. Will the cosmos vanish into oblivion? Will it remain forever in that ultra-compressed state? Or will it "bounce" outward into another phase of expansion, this one too ultimately destined to reverse into a second contraction, a second bounce, and so on without end?

And if such a perpetually oscillating cosmos is possible, is it equally possible that the laws of physics become reset upon each bounce? Could it be that during this cycle of oscillation the electron charge balances that of the proton—but that in the next it will not? Could it be that the previous cycle lasted a mere fraction of a second from creation to destruction because inflation failed to occur? Does the universe teem with life throughout some of its reincarnations but in others remain sterile? Is this particular stage in the infinite history of the cosmos nothing more than a niche—a niche not in space, but in time?

I have no wish to assemble an inconsequential list of the virtues of one particular niche, whether it be geometrical or temporal in nature. How are these two possibilities to be evaluated? The first refers to a property of space we ordinarily never consider: its *connectivity*. In ordinary experience it is always possible to imagine going from here to there, no matter where "here" and "there" may be. The journey may be impractical for one reason or another—"there" may lie on the far side of the Moon, or deep within the Earth—but these are not insurmountable problems. There are ways to overcome them. We say that the space of ordinary experience is connected. The question is whether the universe as a whole has that property.

Experience is no help in answering such a question, neither the ordinary experience of daily life nor the highly refined experience of experiment. By the very nature of what one means by disconnected spaces, evidence could never be found of their existence. Nor do we have any theory to guide us. Einstein's general relativity is a theory of the geometry of space, but questions concerning connectivity belong not to geometry but to a branch of mathematics known as topology. Unfortunately, relativity has nothing to say about topology at all. The closest thing we have to a physical theory of space is utterly silent on this one, crucial question.

We are equally in the dark concerning the second possibility. To evaluate it we would be required to come to an understanding of the Big Bang, that primal moment of infinite compression and temperature from which the present expansion of the cosmos arose, and into which a contracting cosmos is destined to collapse. What happens when all the matter and energy of the universe are crushed together into a region of literally zero size? Is everything crushed out of existence? Or do matter and energy manage in some incomprehensible manner to survive? No one has the slightest idea.

Vapor piles upon vapor, enigma upon enigma.

If the universe was created, were the laws that govern it created too? Or did these principles somehow exist prior to the universe? If the world can be re-created in a second Big Bang, need its laws be re-created in their old familiar form? If multiple universes actually exist, must their laws be identical to our own? Or is the very notion of an alternate reality sheer nonsense? After all, reality is what exists; everything else does not.

Some things seem certain. It makes no sense to consider an alternate reality containing married bachelors. Nor could we postulate a cosmos in which 7 equaled 8. Married bachelors do not exist for reasons involving logic, not physics: What one means by *married* is inconsistent with what one means by *bachelor*. Seven does not equal 8 because 8 is defined to be the successor to 7. To discuss a reality in which such principles were violated would simply be a misuse of language—muddled thought.

But the principles of physics seem different. They do not appear to carry the same weight of inevitability about them. Somehow, it seems not so inconceivable that a universe could exist in which stars were packed closely together— after all, plenty of dense star clusters are known. It does not appear to violate any fundamental laws of thought to imagine the strong interaction being yet stronger. While logic and mathematics are necessarily true, it seems the laws of nature just happen to be true. To this extent, perhaps it is conceivable that alternate realities, each with its own laws, could exist.

But certain physicists have speculated that this view is in

error. They have proposed that in reality the laws of nature possess fully the same degree of inevitability as do those of logic and mathematics. Surely, they have argued, we do not yet know all these laws. Surely some we think we know are false. Perhaps far in the future the day will come when the business of science is completed, and the correct, ultimate laws will be known. And on that day, it may be an extraordinary realization will dawn: that *the true laws of nature are the only mutually consistent ones possible.* Perhaps every other conceivable set will turn out to possess inner contradictions.

If so, all need would vanish for my program of research aimed at elucidating "what if" physics had been different. So too with the notion of alternate sets of laws pertaining to other universes. Reality could not have been different. The laws of other universes must necessarily be those of our own. What we learn in the laboratory of the principles of physics and their fitness for life is universal after all; clouds, wreaths, and vapors evaporate, and certainty returns.

Do you want that certainty? Do you want it right away? I cannot give it to you. Do you want assurance the whole exhausting enterprise adds up to something in the end? I cannot give you that either. There are, however, a few things I can give you.

For surely we are not helpless. Surely there are steps to be taken that will lead us out of this morass. This is the wonderful thing about science, the thing that attracted me to it in the first place and binds me to it still: It provides an objective means for making progress. No matter that my judgment might be clouded by an underlying hope the program of research will lead to such and such a conclusion. It would be no matter even if every scientist in the world were bedeviled by that unacknowledged dream. No matter that scientists are all too human, and prone to every frailty of the race. In spite of all, the scientific method provides a vehicle guaranteed to get you places.

We know what to do next. We need a theory of space, a theory of topology telling us whether disconnected universes are possible. We need a theory of the uttermost mystery that is creation. And we need a theory of natural law—not a theory that explains the operation of the physical world in terms of this law, but a theory telling us how law comes into being in the first place.

We need an understanding of the nature of observation in quantum mechanics, an understanding telling us precisely what an observation is and how it is to be performed. My own opinion—and I stress that it is only an opinion, and not one shared by every physicist—is that this understanding cannot be achieved by operating within the framework of quantum mechanics itself. This framework has already been pushed as far as possible. What we need is a new, more comprehensive theory. And for all its failures and evident inadequacies, Wigner's "mentalist" interpretation and its emphasis on the role played by mind in physics may prove central in guiding us in the construction of this new understanding.

It may be objected that all this is not physics; that what I am calling for is some obscure blend of physics, philosophy, and psychology. Maybe so. But science is perpetually renewing itself by broadening the scope of what it considers acceptable questions. And after all, our separation of knowledge into disciplines is only a human invention. As it is presented to us, the world is one.

In *The Discarded Image*, C. S. Lewis wrote:

> It is not impossible that our own Model
> [of reality] will die a violent death,
> ruthlessly smashed by an unprovoked
> assault of new facts. . . . But I think
> it is more likely to change when, and
> because, far-reaching changes in the
> mental temper of our descendants demand
> that it should. The new Model will not
> be set up without evidence, but the
> evidence will turn up when the inner
> need for it becomes sufficiently great.
> It will be true evidence. But nature
> gives most of her evidence in answer to
> the questions we ask her. Here, as in
> the courts, the character of the

evidence depends on the shape of the
examination, and a good cross-examiner
can do wonders.

It's late. It's late and I'm tired. Time to be getting to bed. As
I stroll around the corner of the house and toward the kitchen
door, it strikes me I ought to be disheartened. So much time
and effort expended, and yet so much more work remaining to
be done! We certainly have our job cut out for us.

But remarkably enough I am not feeling disheartened right
now. As a matter of fact, I'm feeling just the opposite. After
all, it is out of just this state of confusion, flux, and uncertainty that great things develop. Given the nature of the questions we are dealing with, surely if we knew what was going
on, not much would be going on. A smile is on my lips as I
bang the door open.

Psychoanalysts tell us that dreams are expressions of re-
pressed wishes. I'm sure they are right. But in the course of
writing this book a dream came to me, a dream so striking
and so vivid that upon waking I immediately wrote it down.
Nor has the memory of it faded with the passage of time. And
it seems to me that this dream encapsulates perfectly my
state of mind right now, the juncture at which I stand.

In my dream I found myself alone on a great sailboat under way across the ocean. We were out of sight of land. The
sky was a brilliant shade of blue, a blue so perfect and so
deep it took my breath away. Sunlight danced upon the wa-
ters. An immense wind drove a great swell before it. Seagulls
wheeled overhead and uttered their lonely keening cries. We
were on a broad reach, the wind coming in off the right-hand
side, the sails set to leftward.

Each wave of the swell was huge, enormous as it bore
down upon the craft, and flecked with foam as it reared high
above my head. As it rushed toward me I was filled with fear
the boat would swamp, but so well constructed was the craft
on which I found myself that it rode up and over each with
ease.

I could not see forward. My view was obstructed by all
the paraphernalia of a great ship: the immense expanse of the

jib as it strained before the wind, coils of rope and chain, piles of boxes. No matter—the compass would be my guide. Glancing at it I saw we were heading due east, bearing 90 degrees dead on. Suddenly I realized I had not the slightest idea of the craft's destination, nor how distant it might lie. How was I even to know when we finally got there?

Ruminating upon the dream after I awoke, it struck me how appropriate was the image of a sailing craft. A craft is a boat, but *craft* also means "skill," "technique": the accumulated expertise of centuries. That is what carries me forward. But why east? Toward the Old World, scene of the scientific revolution? Toward Eastern religions? Toward the dawn?

None of this concerned me in the dream, however. The boat rushed on. Its speed was dizzying. A mighty hissing swell rose up and bore down upon me. I clung to the tiller. I was reeling with vertigo. The boat lifted proudly up. The wave passed beneath. Not once did I look back.

Appendix

List of Coincidences

Part One of this book developed a list of coincidences—or rather, "coincidences"—upon which the existence of life in the universe depends. Although others are known to science, these, I think, are the most striking. For convenience they are summarized here, grouped according to chapter.

Prologue: The Second Sun.

A remarkable feature of the universe is its emptiness; stars are extraordinarily distant from one another. However, were it not for these vast reaches of empty space, violent collisions between stars would be so frequent as to render the universe uninhabitable. The yet more frequent near-misses would detach planets from orbit about their suns, flinging them off into interstellar space where they would quickly cool to hundreds of degrees below zero.

Chapter 1: The Red Giants.

Life requires chemical elements heavier than hydrogen and helium, both for biochemical reasons and because hydrogen and helium are incapable of forming solids and liquids. These crucial heavy elements are synthesized by nuclear reactions in the cores of stars. For years it was recognized that there was a roadblock standing in the way of these reactions, and it was not known how nature finds its way around that roadblock. Eventually it was discovered that the synthesis proceeds by virtue of two separate resonances between nuclei in the cores of red giant stars; the situation is somewhat analogous to finding a double resonance linking a car, a bicycle, and a truck. Were it not for the coincidence of this double matching, the cosmos would consist solely of hydrogen and helium and could not support life.

Chapter 3: Balance.

The charges of the electron and proton have been measured in the laboratory and have been found to be precisely equal and opposite. Were it not for this fact, the resulting charge imbalance would force every object in the universe—our bodies, trees, planets, suns—to explode violently. The cosmos would consist solely of a uniform and tenuous mixture not so very different from air.

Chapter 4: Blue-White Planet.

Water has a number of striking and unusual properties not shared by any other liquid, properties that make it indispensable to every organism known to science. Its ability to dissolve and transport substances is anomalously great; it plays an essential role in photosynthesis, which in turn is the ultimate source of all food upon the Earth; it is the source of all the oxygen in the atmosphere; its unusually high heat of vaporization enables mammals to regulate their body temperatures by means such as sweating; its ability to store great amounts of heat while undergoing only slight increases in temperature keeps the climate from being bitterly severe; and its peculiar expansion upon freezing is all that prevents most of it from permanently freezing solid.

Chapter 7: The Light of the World.

It is heat from the Sun that keeps the Earth from rapidly cooling down to the near absolute zero of interstellar space, and it is the energy of sunlight that ultimately supplies all the food, and nearly all the energy, employed upon the Earth. But the Sun's existence hangs by a thread. (1) The neutron outweighs the proton by a fraction of a percent; if it did not, neither the Sun nor any other star like it could continue shining for more than a mere few hundred years. (2) A remarkable matching exists between the temperature of the Sun (and other stars) and the absorptive properties of chlorophyll; without this matching, neither photosynthesis nor any other chemical reaction capable of trapping the energy of sunlight could take place. (3) The strong force is just barely strong enough to hold the deuteron together; were it slightly weaker, the deuteron would not exist and stars such as the Sun could not generate energy by nuclear reactions. (4) Conversely, were the strong force slightly stronger, this generation of energy would involve a fuel so ferociously reactive as to be violently unstable.

Chapter 8: Space.

Had space fewer than three dimensions, the complex network of interconnections required for the operation of the nervous system and the flow of blood would be impossible. Had space more than three dimensions, planets could not orbit their suns stably, but would either fall into them or rocket off into the absolute zero of interstellar space.

Chapter 10: The Moment of Creation.

The Big Bang in which the universe originated appears to have been delicately and precisely tuned—tuned in four different ways: (1) Its density was adjusted to have a certain value, the so-called critical density; had it not been, the cosmos would have recollapsed, winking out of existence, the instant after it was created. Furthermore, the Big Bang was (2) perfectly smooth and (3) of a perfectly uniform temperature; had it not been, the subsequent evolution would have led to a state consisting not of stars with their life-giving light and warmth, nor of planets upon which life might flourish, but solely of giant black holes wandering through otherwise empty space. (4) Finally, the apparent symmetry of matter with respect to antimatter implies that the Big Bang's content of each ought to have been symmetrical; had this been so, however, the subsequent evolution would have led to a state containing no matter of any sort. In fact the content of matter was "tuned" to be very slightly greater than that of antimatter.

Glossary

Annihilation Reaction.
Mutual annihilation of a particle and an antiparticle. Their mass is transformed into energy.

Antimatter.
Matter made of antiparticles.

Antiparticle.
To each subatomic particle there corresponds an antiparticle, opposite in charge but identical in every other respect. When particle and antiparticle meet, both are destroyed in an annihilation reaction.

Axon.
Part of a nerve cell. The nerve impulse flows from the cell proper down the axon and then to other cells.

Baryon.

A heavy subatomic particle, such as a neutron or proton.

Dendrite.

Part of a nerve cell. The nerve impulse flows from the dendrite into the cell proper.

Deuteron.

The nucleus of heavy hydrogen, consisting of one proton and one neutron.

Electric (or Electromagnetic) Force.

One of the four fundamental forces of nature. It acts between charged objects: Like charges repel one another while unlike charges attract one another.

Electron.

A subatomic particle. The electron carries negative electric charge and is found in the outskirts of atoms.

Galaxy.

A vast assemblage of stars, containing up to hundreds of billions of them. Galaxies can be either disk-shaped and shot through with winding spiral arms, or elliptical in shape. Ours is spiral and is known as the Milky Way Galaxy.

Grand Unified Theories (GUTS).

Theories uniting the strong, weak, and electromagnetic forces.

Gravitational Force.

One of the four fundamental forces of nature, gravitation is the tendency of matter to attract other matter. It is the only force not included in the grand unified theories.

Heavy Hydrogen.

Hydrogen whose nucleus is a deuteron, as opposed to the normal form whose nucleus is a single proton.

Interstellar Cloud.
A cloud of highly rarefied gas found in interstellar space.
New stars and planets are thought to form from such clouds
by a process of condensation.

Milky Way.
The disk of our galaxy (known as the Milky Way Galaxy)
seen edge-on.

Natural Theology.
An outmoded nineteenth-century theory, which held that the
high degree of organization evident in living organisms owed
its existence to God.

Neuron.
A nerve cell.

Neutron.
A subatomic particle. The neutron carries no charge, and is
one of the two constituents of the atomic nucleus (the other
being the proton).

Nucleus.
The central, heavy part of an atom, consisting of neutrons
and protons. The nucleus is positively charged, and so attracts to it negatively charged electrons to form an atom.

Photon.
The particle of light. It has no mass and no charge.

Photosynthesis.
The chemical reaction whereby plants trap the energy of sunlight and convert it into a biologically useful form.

Proton.
A subatomic particle. The proton carries positive electric
charge, and is one of the two constituents of the atomic nucleus (the other being the neutron).

Quark.

The elementary particle of which subatomic particles such as neutrons and protons are composed.

Strong Force.

One of the four fundamental forces of nature. It acts between quarks and certain subatomic particles and binds together the atomic nucleus.

Weak Force.

One of the four fundamental forces of nature. It acts between certain subatomic particles and is responsible for their decay.

Index

About the Author

Astrophysicist and author George Greenstein received his B.S. from Stanford and Ph.D. from Yale, both in physics. He pursued a growing interest in astronomy through postgraduate studies at Yeshiva and Princeton universities, and since 1971 has been at Amherst College and the Five College Astronomy Department. From 1981 to 1984 he was chairman of the Five College Astronomy Department, and at present he is professor of astronomy at Amherst College. Professor Greenstein is the author of numerous technical articles on astronomy and astrophysics, and is a member of the International Astronomical Union, the American Astronomical Society, and the Authors Guild. His previous book, *Frozen Star*, was awarded the Phi Beta Kappa Award in Science for 1984 and the American Institute of Physics–United States Steel Foundation 1984 Science-Writing Award.